The
Economic
Feasibility
of
Recycling

The Economic Feasibility of Recycling

A Case Study of Plastic Wastes

T. Randall Curlee

PRAEGER

New York
Westport, Connecticut
London

Permissions

Table 5.4. Cost data other than landfill and material cost gathered from Goodyear Tire and Rubber Company. "Recycling Polyester Bottles." *Cleartuf Facts*, CT-17. n.d. Akron, Ohio. Used with permission.

Tables B.1–B.11. Historical data and numerical values gathered from Society of the Plastics Industry. *Facts and Figures of the U.S. Plastics Industry*. Various issues. Used with permission.

Tables C.1–C.2. Information gathered primarily from Scrap Age. "Exclusive Updated Survey of Automobile Shredding." October 1980. Pp. 91–98. Used with permission.

Figures 4.1–4.3. Historical data gathered from Society of the Plastics Industry. *Facts and Figures of the U.S. Plastics Industry*. 1984–1985. Used with permission.

Library of Congress Cataloging-in-Publication Data

Curlee, T. Randall.
  The economic feasibility of recycling.

  Bibliography: p.
  Includes index.
  1. Plastic scrap—United States—Recycling.
I. Title.
HD9661.U62C87 1986   668.4'192   86-22675
ISBN 0-275-92376-2 (alk. paper)

Library of Congress Catalog Card Number: 86-22675
ISBN: 0-275-92376-2

First published in 1986

Praeger Publishers, 521 Fifth Avenue, New York, NY 10175
A division of Greenwood Press, Inc.

Printed in the United States of America

The paper used in this book complies with the Permanent Paper Standard issued by the National Information Standards Organization (Z39.48-1984).

10 9 8 7 6 5 4 3 2 1

To my family

Contents

Figures

Tables

Acknowledgments

In the course of researching and writing this book helpful assistance was received from numerous individuals whom I would like to acknowledge and thank. Special thanks go to Joseph Carpenter of the Metals and Ceramics Division at Oak Ridge National Laboratory (ORNL). His invaluable technical expertise, managerial skills, and continued commitment to this work not only made the completion of this book possible, but also made the endeavor a rewarding and even enjoyable undertaking. Financial support for this work through the U.S. Department of Energy's Energy Conversion and Utilization Technologies Program (ECUT) is greatly appreciated.

The managerial support for this project provided by Bill Fulkerson, Tom Wilbanks, and Bob Shelton within the author's own Energy Division at ORNL is gratefully acknowledged. I owe Bob Shelton particular gratitude for his encouragement to publish this work from its inception and his assistance in the somewhat tedious task of negotiating with the ORNL book-writing department.

I am indebted to Al Ekkebus of the Central Research Library at ORNL for his superb assistance in identifying relevant materials for this work and to Opal Russell also of the Central Research Library for her assistance in document retrieval. The retrieval of data for this work was greatly facilitated by Howard Kibbel of the Society of the Plastics Industry and Terry Yaconna of Manning, Selvage and Lee. Special recognition goes to Barbara Leffel, Catherine Woods, Michael Fisher, Michael Esposito, and Mia Crowley of Praeger for their courteous and valuable editorial support. Helpful secretarial support was provided by Lil Cochran and is gratefully acknowledged.

The development of this book has benefited enormously from discussions with individuals who have been instrumental in the development and evaluation of processes to recycle plastics. These individuals include Michael Bever of the Massachusetts Institute of Technology; Michael Curry, Lincoln Hawkins, and Albert Spaak of the Plastics Institute of America; Sidney Rankin of the Plastics Recycling Institute; Jack Milgrom of Walden Research; John Lawrence, formerly of the Society of the Plastics Industry; and Jacob Leidner of the Ontario Research Foundation. I am particularly indebted to the work of Jacob Leidner, especially his excellent 1981 book, *Plastics Waste*, which provides a detailed discussion of various technical issues relevant to plastics recycling. That work was instrumental in my decision to address the economic and institutional factors that will influence the overall viability of recycling technologies. I am also indebted to Dave Bjornstad of ORNL, Lincoln Hawkins, and Jack Milgrom for their helpful and constructive reviews of an earlier draft of this work. I am, of course, solely responsible for any of the book's remaining shortcomings and inadequacies.

My greatest and most varied debt is, however, to my family. My wife, Deb, not only provided moral support during the research and writing of the book, but also supplied valuable and constructive criticisms about the approach used and editorial suggestions on an earlier version of the work. Her willingness to listen to and question my arguments furnished an invaluable method to develop ideas and organizational approaches. And her understanding and encouragement during the sometimes laborious and tedious writing phase made an otherwise arduous task less demanding. My son, Colin, deserves special recognition. He not only sacrificed irreplaceable play time with his dad so that the book might be completed, but also gave purpose and meaning to the whole project with his exuberant hugs and insatiable curiosity. His enthusiasm and sometimes comical perspective often provided a unique and valued diversion. I want to thank my parents for their interest and excitement about, and support of, this work. And to my larger family, I am grateful for their much needed support and encouragement.

The
Economic
Feasibility
of
Recycling

1
Introduction and Overview

INTRODUCTION

A significant amount of work has been done in recent years to address the problems, as well as opportunities, posed by the production and accumulation of plastic wastes. However, the vast majority of work has focused on technological questions and has largely neglected the economic and institutional incentives and barriers that may have a pronounced impact on the degree to which recycling technologies are ultimately adopted in the marketplace. This book addresses the problems and opportunities associated with plastics recycling from an economic perspective, and reviews numerous economic and institutional factors that have not heretofore been studied — factors that may largely determine whether future plastic wastes will be disposed of or recycled.

The increased interest in plastics recycling has arguably resulted from three important trends. First, the production and use of plastic resins in the United States has more than quadrupled during the past two decades. Plastics now compose approximately six percent of the typical municipal waste stream and projections indicate that this percentage will increase during the coming decade. Virtually every market segment now uses plastics in some form as technological innovations result in new plastic resins and composites with widely varying physical and chemical properties. Furthermore, it is generally agreed that most markets will increase their usage of plastic resins in coming years as the properties and cost competitiveness of more conventionally used materials are surpassed by those of existing and forthcoming plastics.

Second, while the relative price of plastic resins and materials has decreased in relation to the prices of metals and paper products during recent years, resin prices have increased dramatically in absolute terms.[1] The price

1

of plastics rose sharply following the drastic petroleum price increases of 1973–74 and 1978–79. (Petroleum and natural gas are, of course, the primary raw materials used in the production of plastic resins.) The value of the plastics entering the waste stream has therefore been perceived to have risen in accordance with increases in the prices of virgin resins. Further, as the prices of oil and other traditional energy sources have escalated, the energy content in plastic wastes, which is roughly equivalent to that of coal on a per pound basis, has become increasingly valuable.

Third, growing concern about the quality of the environment has led to more restrictive disposal regulations. While there is disagreement about the extent to which plastics pose environmental problems in waste disposal, there is agreement that waste disposal is, in general, becoming more expensive. From an economic perspective, these escalating disposal costs make plastic waste recycling more viable. And to the extent that plastic wastes may be harmful to the environment, the public has an incentive to reduce the volume of plastics entering the disposal stream.

In response to these trends, numerous studies and research and development projects have been undertaken to address the technological problems posed by the recycling of plastic wastes. A flurry of activity occurred in the early 1970s to develop technologies that could separate plastics from other types of waste and recycle that waste to produce new plastic products, retrieve basic chemicals from the waste, or retrieve the heat content of the waste. Technologies have been developed that utilize plastic wastes to produce products as diverse as fenceposts and fuel oil. In many cases these technological efforts have been successful and several recycle technologies are now being tested or are fully operational. Many claim to be economically viable. Moreover, work continues on new processes that may increase the quality and reduce the cost of products derived from plastic wastes.

However, at this time the technologies that are available to recycle plastics have met little success in penetrating their potential markets, especially in the United States. The quantity of plastic waste that is currently recycled is minuscule when compared with the total quantity of plastic waste currently being produced. Further, with the exception of a few isolated market segments, there is little current movement toward the adoption of practices to recycle plastics on a large scale. There is an increasing recognition that plastics recycling is an extremely complicated issue, not only from a technological perspective but also from economic and institutional perspectives.

The primary purpose of this book is to identify and study the numerous economic and institutional incentives and barriers that impact on the decision to recycle or dispose of plastic wastes. These incentives and barriers are addressed from the perspective of individuals and firms that must make a decision to either recycle or dispose of plastic wastes and from the perspective of the general public that must through legislation and regulations encourage recycling, discourage recycling, or be neutral on the issue. An overview of the purposes, scopes, and limitations of the analyses presented in subsequent chapters that address these and other issues is the topic of the following section.

OVERVIEW OF SUBSEQUENT CHAPTERS

Chapter 2. While it is not the purpose of this book to present a detailed or definitive discussion of the various technological issues that are relevant to plastics recycling, it is imperative that a minimal understanding of the technological issues be obtained in order to understand the intricacies of the economic and institutional issues that may impact on the decision to recycle. Chapter 2 contains a relatively nontechnical discussion of these technological issues. There are three focuses of the discussion: a) the various types of plastics and how the characteristics of different resins impact on their recyclability; b) the different types or groups of technologies that can be used to dispose of or recycle plastic wastes and some specific examples of technologies within each technology group; and c) the environmental impacts of disposing or recycling plastics by using the various available technologies.

One of the major, if not the major, technological problem to be faced in recycling plastic wastes is the diversity of the physical and chemical characteristics of the numerous resins that are called "plastic". There are two main types of plastics — thermoplastics, which can be repeatedly reformed by softening or melting under heat and pressure, and thermosets, which cannot be remelted once their interlinking molecular bonds are formed. Obviously, the characteristics of thermosetting resins make those resins much more difficult to recycle than thermoplastic resins. Further, within each major group of plastic resins, the physical and chemical characteristics of the different resins vary widely. (These specific characteristics are addressed in more detail in Chapter 2 and in Appendix A, which summarizes the characteristics and typical uses of some of the major thermoplastic and thermoset plastic resins.) While one thermoplastic may melt at, for example, 150°F, another thermoplastic may have a melting temperature in excess of 500°F. Moreover, it is difficult to predict how the physical characteristics of a mixture of different resins will vary from the characteristics of the resins when used individually. Therefore, in no sense can plastic resins be thought of as being uniform. Different resins vary in characteristics and properties as much as, for example, steel differs from aluminum. These heterogeneous characteristics of resins severely limit the types of recycling that are technically feasible and greatly complicate the discussion of the economic feasibility of plastics recycling.

The recycling of plastic wastes is further complicated in that a plastic resin in one waste stream is not identical to the same resin in another waste stream. As will become clearer in Chapter 2, a major technical constraint to the recycling of plastic wastes is the form in which that waste enters the waste stream. While significant technical progress has been made toward separating plastics from other wastes, it is generally recognized that once a plastic enters the municipal waste stream, it is technically very difficult and not economically viable to separate that plastic from other municipal wastes. Therefore, a major key to recycling plastic waste, at least in a relatively uncontaminated form, is our ability to collect and process that waste before it enters the municipal waste stream. Obviously this ability depends on the product in which the waste resin appears and how that product is typically disposed of. The collection possibilities and therefore the recycling potential for a thermoplastic waste in, for example, the form of a returnable soft drink bottle are vastly different from the possibilities for the same resin when it appears in the form of consumer product packaging, which is extremely difficult to divert from the municipal waste stream.

Another technical complication that distinguishes the issue of plastics recycling from other widely recycled materials — such as steel, aluminum, and copper — is the various ways plastics can be recycled. Unlike metal recyclers that take metal scrap in a contaminated form and basically reduce the contamination to a level acceptable for the intended use, plastics recyclers have several alternatives. Recycling technologies are usually discussed in one of four major categories. Primary recycling is the processing of waste into a product with characteristics similar to those of the original product. The plastics industry has historically recycled much of the waste that occurs during resin production, fabrication, conversion, and product assembly in a primary sense. However, the primary recycling of plastic wastes does not allow for any significant waste contamination and therefore is not applicable for more contaminated manufacturing wastes (sometimes referred to as manufacturing nuisance plastics) and virtually all types of postconsumer plastic wastes. For the most part, this book is not concerned with the primary recycling of clean manufacturing waste, but rather focuses on those manufacturing and postconsumer plastic wastes that have historically been disposed of by landfill or incineration.

Secondary recycling is the processing of plastics waste into products with characteristics that are inferior to those of the original product. As is discussed in some detail in Chapter 2, there are numerous technologies that are available to recycle thermoplastic wastes into products such as drainage pipes and construction materials similar to wood products. The use of these technologies is predominantly limited by the percentage of the total waste that is composed of thermoplastics. This technological constraint reemphasizes the importance of collecting plastic wastes independently of other municipal wastes or having the capability to economically separate plastics from other waste materials. There is also some discussion in the general literature of recycling thermosetting wastes in a secondary sense by grinding and using the thermosets as a filler material with virgin resins.

A third recycling category that has received a great deal of attention is tertiary recycling, which attempts to recover basic chemicals and fuels from waste resins. Processes such as pyrolysis and hydrolysis fall into this category. Several tertiary technologies have been developed and tested, and in some cases have been used commercially. One currently used technology produces No.6 and No.2 fuel oil from polypropylene manufacturing waste that has historically been disposed of by landfill.

The fourth technology category is quaternary recycling, which retrieves the heat content of plastic wastes by incineration. Depending on the specific resin, plastic wastes can be a good fuel source. Most resins have heating values of at least 12,000 Btus per pound, which is roughly equivalent to that of coal. Potential problems associated with this form of recycling include the production of toxic fumes and undesirable chemical reactions that can damage the heat-retrieving incinerators.

An important point to make with regard to tertiary and quaternary recycling is that plastics do not necessarily have to be segregated from other municipal wastes in order for these technologies to be used. There are several operations in the United States that currently utilize, or have utilized, municipal waste—inclusive of plastics—in a tertiary or quaternary sense. The restrictive collection and separation requirements of primary and secondary recycling do not therefore necessarily apply to tertiary and quaternary recycling.

Another important technological issue, especially from a social or public perspective, is the environmental damage that may result from the incineration or landfill of plastic waste (the usual disposal methods), as well as the potential environmental damage that may be caused by different forms of recycling. As is discussed in Chapter 2, there is no general agreement about the environmental impacts of the various methods of disposing of plastics. Some argue that the production of toxic fumes from plastics incineration, with or without heat recovery, makes that form of disposal or recycling socially unacceptable. Others argue that the pollutants are not severe, especially when appropriate technological controls are taken. The environmental impacts of plastics in landfills are equally controversial. Some argue that the nonbiodegradability of most resins renders plastics unacceptable for landfill, while others argue that this very characteristic can provide structural support to the landfill over the longer term.

What is clear, however, is that the recycling of plastic wastes in whatever form and through whatever technology will not completely eliminate the potential environmental problems posed by plastic wastes. Quaternary recycling will pose risks similar to those of incineration without heat recovery. Tertiary recycling will present environmental problems similar to basic chemical production. And the secondary recycling of plastics will in most cases require some degree of waste separation, thus reducing but not totally eliminating the waste that must be disposed of by landfill or incineration. Further, it is generally recognized that some degree of degradation of the thermoplastic's chemical and physical properties occurs in secondary recycling. This degradation will eventually force the plastic being recycled to enter a disposal stream or be recycled in a tertiary or quaternary sense. Secondary recycling therefore only prolongs the life of the resin, rather than allowing for the resin's infinite recycling.

When compared to other types of recycling, e.g., metals and paper, the recycling of plastics is very complicated from a technological perspective, which, in turn, complicates the issues that must be addressed from an economics or institutional perspective. Different resins vary significantly in their physical and chemical properties. The numerous products in which postconsumer plastic wastes appear further limits the potential for recycling. In addition, the potential recycler faces a host of recycling possibilities, each offering certain pros and cons. Finally, the environmental impacts of disposal and recycling depend on the technology adopted and are not generally well understood even within a particular technology class. Chapter 2 addresses these technological issues in more detail and discusses their relationships to the economic and institutional issues that the public and private sectors will face in coming years.

Chapter 3 discusses in a conceptual sense the economic and institutional issues that impact on the decision to recycle or dispose of plastic wastes. The discussion is presented from two perspectives: that of individuals and private firms that are involved in the production, disposal, and recycling of plastic wastes; and that of the public sector that must through legislation and regulations voice an opinion on the issue.

From the perspective of the private firm that produces plastic wastes, there is an incentive to recycle when the direct net costs of recycling are less than the direct costs of disposal. In other words, for the firm that produces waste, recycling opportunities need not be economically attractive in themselves, but simply must be more attractive than the alternative of disposal.

The picture is somewhat clouded, however, by what may be termed potential institutional problems. For example, while a recycling technology may be potentially less costly than disposal at some minimal level of operation, an individual firm may not produce large enough quantities of waste for the recycling technology to be economically viable. In this case the waste producer is dependent on the development of markets that can utilize waste from several producers in operations of sufficiently large scale. Other similar problems can be foreseen and are discussed in Chapter 3.

The formation of markets for plastics recycling is, however, often stifled by the great uncertainties faced by entrants into that market. Uncertainties arise from various sources. The quantities, types, and qualities of plastic waste available for recycling are often not known, or are believed to vary significantly over time. A potential investor in a recycling operation thus faces the risk that the required waste will not be available for recycling. The technological uncertainties associated with the new and relatively untried recycling processes pose another significant source of risk. Will the technologies perform as they are designed, and will the resulting products have satisfactory properties? Further, to what extent will the products produced from recycled plastics be accepted in the market place? In the vast majority of cases, recycled plastics will not compete with virgin resins, but rather will compete with materials such as wood or metals. The recycler must face the potential that purchasers of products made from recycled plastics will be biased against those products because they feel the products are in some way inferior. These and other uncertainties may force potential investors in recycling operations to demand a significant risk premium for those operations. Depending on the firm's degree of risk aversion, the potential uncertainties associated with recycling may, in fact, force some firms to reject recycling and opt for disposal even though the expected costs of recycling are less than the expected costs of disposal.

From the perspective of society, two sets of economic and institutional issues arise. First, should the public sector through regulations and legislation encourage the private sector to recycle plastics rather than dispose of plastic wastes by incineration or landfill? Second, to what extent should the public sector make efforts to recycle postconsumer plastics that appear in the municipal waste stream?

In the first case, the proper public sector involvement is dictated by the degree to which the private market fails to operate optimally from a social perspective. There have been numerous arguments why governments should encourage recycling in general. However, empirical studies of these arguments have largely been limited to the recycling of metals and paper. The fundamental argument is that the disposal of waste imposes external costs of various types on society — costs that are not directly realized by the individual disposers or processors of that waste. External environmental costs are most often discussed. It is argued that government intervention will be required to force the cost of disposal that is realized by the waste producer or processor to reflect all external, as well as internal, costs. Other arguments focus on potential inequities between the competitiveness of recycled materials and virgin materials. It is often alleged that the inequities have resulted from government involvement and therefore require additional government actions to correct the problems. These various arguments are discussed in detail in Chapter 3 and applied to the case of plastic wastes. It is suggested that while some of the generic arguments for the government promotion of recycling do apply to plastics, other arguments do not logically follow because of technological and institutional differences between plastics and other recycled materials.

In the second case, many of the arguments for the first case also apply. However, there are sufficient technical and institutional differences between plastics in municipal waste and plastics from sources that have not entered the general waste stream to draw a distinction. Recall that once a plastic enters the municipal waste stream it is very difficult to separate that plastic from other materials. This problem significantly limits the recycling alternatives and renders some types of recycling economically unattractive. As is discussed in more detail in Chapter 3, the economic and institutional issues relevant to postconsumer waste differ depending on whether the waste is difficult or relatively easy to divert from the municipal waste stream.

Chapter 4. One of the principal uncertainties facing both private and public decision makers is the production of plastic wastes in future years. Chapter 4 contains estimates and projections of manufacturing and post-consumer plastic wastes for the years 1984, 1990, and 1995. In addition, the chapter contains a discussion of recent trends in the use of resins by major market categories. Projections of total resin use in several U.S. markets are given through the year 1995.

As discussed in the overview of Chapter 2, two crucial elements in determining the recyclability of future plastic wastes are the quantities of, and forms in which, particular plastic resins will enter the waste stream. Wastes are divided into two main categories, manufacturing nuisance plastic wastes and postconsumer plastic wastes. Using projections of future resin production and historical estimates of the percentages of plastic resins that become waste at various processing steps, estimates of future manufacturing nuisance plastics are made. The waste projections are disaggregated according to several major thermoplastic and thermosetting resin types. It is projected that manufacturing nuisance plastics in the U.S. will increase in quantity from the estimated 2.8 billion pounds in 1984 to about 4.0 billion pounds in 1995. About 82 percent of those wastes are projected to be thermoplastics. The largest single waste-producing manufacturing step is fabrication, which produces about 32 percent of the total.

Postconsumer waste estimates and projections are presented for nine major market categories and for fifteen major resin types within each category. The particular methodology employed to produce these estimates and projections differs depending on the market category because of constraints on data availability. However, the basic methodology depends on historical information about the usage of plastics in different products and information about the average lives of those goods. Total postconsumer plastic wastes are projected to increase from the estimated 29.6 billion pounds in 1984 to 43.4 billion pounds in 1995. Thermoplastics are estimated to have composed about

87 percent of all postconsumer wastes in 1984 and are projected to increase slightly to about 88 percent in 1995. Packaging is by far the largest single contributor to postconsumer plastic wastes, in excess of 42 percent of the total for all years. Plastic waste from the building and construction sector is projected to grow fastest of all product categories. Various postconsumer product categories are combined with industrial nuisance plastics to suggest the types, quantities, and qualities of plastic wastes, which then will be difficult or relatively easy to divert from the municipal waste stream.

To assist in the waste projections and give an indication of resin usage by product category, projections of total U.S. resin production and the use of total resins in major product categories are given. Significant growth in resin usage is projected for the packaging and construction sectors. Total U.S. resin production is expected to increase from the 42.7 billion pounds produced in 1983 to about 63.8 billion pounds in 1995.

Chapter 5 abstracts from the total costs of plastics recycling and disposal (inclusive of external costs) and focuses on published estimates of the direct expected costs and revenues associated with particular technologies. Unfortunately in many cases the technologies available for plastics recycling have not been used extensively or long enough to provide reliable cost and revenue estimates. In many cases the only information on the economic viability of the processes comes from the developers of the processes. In other cases, the information about a particular process varies significantly from source to source. Moreover, many estimates are somewhat dated or incorporate restrictive assumptions that make comparisons with other technologies difficult.

Nevertheless, information on the direct costs of disposal and the expected costs and revenues from the available and proposed recycling technologies is of crucial importance to both public decision makers and private firms contemplating recycling operations. Chapter 5 therefore reviews and summarizes several published estimates of the costs of different disposal processes and the costs and revenues associated with different recycling technologies. Disposal alternatives include several types of landfill and incineration. Various secondary, tertiary, and quaternary recycling technologies are considered. In all cases, the published estimates are converted to constant dollars to allow limited comparisons among the numerous options. While at this time the availability of data is quite limited and the quality of some data is in question, the available data indicate that in many cases plastics recycling is superior to disposal on a direct expected cost basis. Further, the calculations suggest that some tertiary and quaternary recycling processes are competitive with many secondary processes.

Chapter 6 contains detailed examinations of three market sectors where secondary, tertiary, and/or quaternary plastic waste recycling is currently in progress or has a good chance of being implemented in future years. The sectors include carbonated beverage bottles that are manufactured from PET (polyethylene terephthalate), the electrical and electronics sector (with a particular focus on telephone equipment), and shredder residue from automobile recycling operations. These selected market sectors do not represent all the areas where recycling is currently taking place or where good possibilities for recycling exist. These sectors are, however, a representative sample of the types of waste streams that lend themselves to plastics recycling as a relatively uncontaminated waste, rather than as a part of the municipal waste stream.

In the case of each market sector, the chapter examines the specific economic, institutional, and technical incentives and barriers that have encouraged and discouraged recycling. The intent of this exercise is to identify those conditions that facilitate recycling as a viable alternative to disposal.

While numerous specific conditions are discussed with respect to each market segment, an overriding precondition for plastics recycling in a relatively uncontaminated form is the incentive and ability to collect the waste plastics prior to their entry into the municipal waste stream. In the case of soft drink bottles, collection outside of the municipal waste stream is greatly facilitated by bottle deposit laws in several of the more populous states. In the cases of electrical and electronic equipment and automobiles, collection is facilitated by well established recycling markets that have historically recycled the metallic portions of these products; a large percentage of the remaining residue is often composed of plastics.

Chapter 7. Drawing on the major conclusions from previous chapters, the final chapter attempts to answer several general questions about plastics recycling. In this process, the chapter reviews the major economic and institutional incentives and barriers that will influence our future decision to recycle or dispose of the growing quantities of plastic wastes. The overriding conclusion of this book is that, given the current economic and institutional incentives and barriers and the current state of recycling technologies, the recycling of plastics as a relatively uncontaminated waste can be expected to grow in future years in certain narrowly defined product areas. However, the recycling of plastics as a segregated waste is not expected to occur to the extent we currently observe in the major metals, e.g., steel, aluminum, and copper. Fundamental economic, institutional, and technological constraints will limit the recycling of plastics outside of the municipal waste stream to a small percentage of the total plastic wastes projected to be produced during the coming decade.

Recycling within the municipal waste stream is, however, currently on the rise. That trend is expected to continue as environmental restrictions elevate the cost of landfill and thus make the tertiary and quaternary recycling of municipal waste — inclusive of plastics — an attractive economic alternative. These trends will be particularly pronounced in areas with large populations.

NOTE

1. In 1983 the producer price indexes for metals and metal products, paper and allied products, and plastic resins and materials were 307.1, 297.7, and 290.2, respectively. This indicates that the price of plastics has increased relatively less than the prices of competing materials since the base year of 1967.

2
Technological Issues

INTRODUCTION

It is not the purpose of this book to provide a detailed or definitive discussion of the numerous technological issues that affect the feasibility of plastics recycling. However, in order to understand the economic and institutional incentives and barriers that influence the decision to recycle, one must first have a general knowledge of the technological barriers that limit the recycling of plastic wastes. The major purpose of this chapter is therefore to present a relatively nontechnical discussion of the technological parameters that impact on recycling. For the most part, the chapter abstracts from the economic issues discussed in other chapters of this book. However, where applicable the chapter discusses the importance of a technical barrier to the overall economic feasibility of plastics recycling. A secondary purpose is to refer the interested reader to sources that give more technical discussions of the issues presented here.

There are three main forces of the chapter. The first examines how the physical and chemical characteristics of different plastic resins in the waste stream impact the potential recyclability of those resins. In no way can plastics be considered a homogeneous material. Plastic resins differ in their characteristics as much as, for example, wood differs from steel. Further, the recyclability of a plastic waste will depend on the way the waste is typically collected, which, in turn, largely determines the degree to which plastic wastes are contaminated with other materials. For example, a returnable plastic beverage bottle, which is typically collected outside of the municipal waste stream, is vastly different from plastic consumer product packaging, which almost always is collected as part of the municipal waste stream.

A second focus of the chapter is on the technical characteristics of the currently available recycling and disposal processes. Different recycling processes are applicable to different types of plastics with varying levels of waste contamination. The technologies also differ in terms of the products produced from the waste materials.

A third focus is on the potential environmental consequences of disposing plastic wastes, as well as recycling those wastes. One of the major arguments given for government programs to encourage recycling concerns the environmental damage that waste disposal entails. Several technical studies that have addressed these impacts are reviewed and summarized. What has been emphasized less in the literature is the potential environmental damage associated with plastics recycling. While some forms of recycling may significantly reduce environmental damage, recycling does not eliminate the environmental problems posed by the rapid growth in the production and disposal of plastic products. Many of the current and proposed recycling processes will themselves emit their own significant levels of pollutants.

AN OVERVIEW OF TECHNOLOGICAL ISSUES

Before we examine the technological issues at a disaggregated level, it is advantageous to view the parts of the problem from a broader perspective. Figure 2.1 presents a simple flow diagram that may be helpful in introducing the technological issues to be considered in the recycling and disposal of plastic wastes.

There are two main sources of plastics wastes: manufacturers (i.e., resin producers, fabricators, converters, assemblers, packagers, and distributors), and consumers of plastic products. Manufacturers have recycled much of their uncontaminated waste for some time and this is now considered standard practice. However, there is a significant portion of manufacturing waste that is contaminated with other materials or is otherwise considered unsuitable for recycling. These wastes, which are often referred to as manufacturing nuisance plastics, have historically been disposed of by landfill or incineration, and it is with this portion of manufacturing waste that this study is predominantly concerned. Historically, consumers have not recycled a significant portion of their plastic wastes. While a few postconsumer plastic materials have been diverted from the municipal waste stream because of, for example, bottle deposit laws, the vast majority of postconsumer plastics have been disposed of with other municipal wastes.[1]

The first step toward either disposal or recycling is waste collection. As is discussed in more detail below, the way in which waste is collected is very important because it is at this point that waste contamination can be avoided. If plastic wastes can be collected outside of the municipal waste stream, costly

FIGURE 2.1. Flow Diagram of Plastic Waste Disposal or Recycling

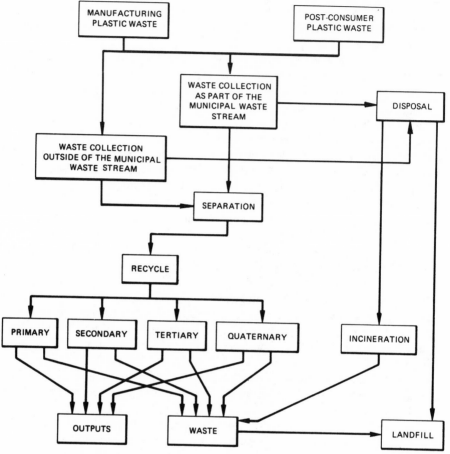

and technologically difficult separation processes required for most forms of recycling may not be necessary.

Once the waste is collected, the plastics can enter the conventional disposal stream (usually incineration or landfill) or be diverted to a recycling stream. In the case of landfill, the technological problems are not usually considered severe, at least with nontoxic and relatively inert wastes such as plastics. The environmental problems of disposing plastics in landfills are more controversial and are either severe or insignificant, depending on the source cited. A consensus problem with landfill is, however, the decline in available land for landfill operations, especially around densely populated areas.

In the case of incineration, plastic wastes pose environmental problems because of toxic fumes that may result from the burning of some resins. Technological problems may also arise because the burning of some plastic materials may damage the incinerator. Further, in the case of incineration, noncombustible waste will be produced that is normally disposed by landfill. The percentage of incoming waste that remains after incineration can be significant. If plastics are burned with other municipal wastes, the residue in weight will measure approximately 20 to 30 percent of the incoming waste. [See Diaz, Savage, and Golueke (1982) for details.]

While there are less costly disposal options, such as open dumping, those options can be ruled out because of their excessive environmental costs. Moreover, the Resource Conservation and Recovery Act of 1976 requires an eventual prohibition on the open dumping of solid wastes.

If recycling is selected, the first step is usually the separation of plastics from other waste materials. While the degree to which plastics must be separated depends on the particular recycling technology, separation, especially for plastics in the municipal waste stream, has been one of the most difficult technological and economic problems to overcome. Numerous technologies have been developed to separate wastes of different types with varying degrees of success. Work continues on this crucial technological hurdle to develop separation processes that are both technologically and economically sound.

Recycling technologies are usually divided into four types: primary, secondary, tertiary, and quaternary. Primary recycling is the processing of a waste into a product with characteristics similar to those of the original product. The recycling of relatively uncontaminated waste plastics, which has historically taken place in the manufacturing sector, is an example of primary recycling. However, little hope exists from both technological and economic perspectives for the primary recycling of plastic wastes that have been contaminated with other waste materials. The current state of separation technologies does not permit the economical separation of plastics from a contaminated waste stream.

Secondary recycling is the processing of waste plastics into materials that have characteristics that are less demanding than those of the original plastic products. Some manufacturing and postconsumer wastes currently enter secondary recycling streams, which allow higher contamination levels than primary recycling. Secondary processes usually produce products such as fenceposts and other bulky items that usually substitute for wood, concrete, or metal.

Tertiary recycling involves the production of basic chemicals and fuels from plastic waste as part of the municipal waste stream or as a segregated waste. Pyrolysis and hydrolysis are examples of these processes. Tertiary recycling is currently taking place and is generating a great deal of interest mainly because relatively high levels of waste contamination can be accommodated.

A final type of recycling is quaternary recycling, which attempts to retrieve the energy content of the waste by burning. In the case of plastics, this type of recycling can be very beneficial because of the high heat content of most plastics. While heating values of different resins vary from almost zero to over 20,000 Btus per pound, the average plastic contains about 12,000 Btus, or about the same as anthracite coal on a per pound basis.

Each of these recycling types has its own associated outputs that must compete with other materials and waste products that must be disposed of. Further, each process has its own set of good and bad points from technological, environmental, and economic perspectives. The following sections of this chapter address the technological and environmental problems posed by waste separation and by each recycling and disposal type. The economic perspective is addressed in the remaining chapters of this book.

PLASTIC TYPES

Perhaps the most difficult technical problem posed by plastics recycling is coping with the widely varying physical and chemical characteristics of different plastic resins. The diverse properties of different resins will in some cases mean the recycling possibilities for the resin are virtually nil and in other cases may mean recycling is relatively simple.

There are two main types of plastic resins: thermoplastics and thermosets. Thermoplastics differ from thermosets in that they can be repeatedly softened and hardened by heating and cooling without seriously damaging the properties of the resins. Thermosets, however, cannot be heated and reformed into new products without destroying the resin's physical and chemical properties. In simple terms, the difference between thermoplastics and thermosets can be explained by the presence or absense of cross-linking in the molecular structures that form the resins. All plastics are composed of long molecular chains that may or may not be attached to one another. In the case of thermoplastics, the molecular chains are not attached. Therefore, as the temperature of the resin increases, the additional energy allows the molecules to move unrestrained. When cooled the molecular structure becomes rigid. However, in the case of thermosets, the molecular structures are cross-linked, which prevents the movement of the molecular chains beyond a certain point. This characteristic, while making the resins stronger when subjected to higher temperatures, also makes thermosets very difficult to recycle. Once the molecular structures of thermosets are formed, those resins, unlike thermoplastics, cannot be heated and reshaped into new products. Figure 2.2 gives a simple graphic display of the molecular structures of thermoplastics and thermosets.[2]

On a weight basis, thermosets have in recent years accounted for about 15 percent of the total domestic production of plastic resins. Common ther-

**FIGURE 2.2.   Molecular Structures of Thermoplastics and Thermo-
sets**

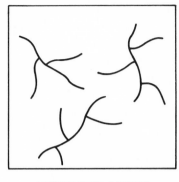 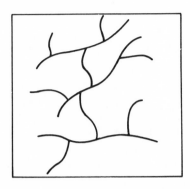

THERMOPLASTICS   **THERMOSETS**

mosetting resins include epoxy, melamines, unsaturated polyesters, and urea. Common thermoplastics include nylons, polyethylene, polyesters, polypropylene, polystyrene, polyvinyl acetate, and polyvinyl chloride. Appendix A of this book gives a glossary of characteristics and common uses of selected thermoplastic and thermosetting resins.

SEPARATION

The key technological constraint that makes the collection of plastic wastes outside the municipal waste stream so important is separation. While a great deal of effort has been devoted to the separation of plastics from other materials, significant technical problems must be overcome before the separation of plastics from the municipal waste stream can be considered a viable economic option. This section reviews recent technical developments in the separation of plastic wastes and discusses some of the technical problems to be overcome.[3]

While the use and disposal of plastic goods has increased rapidly in recent years, plastics still do not compose a large percentage of total municipal wastes. Estimates of the actual percentage vary depending on the source. Caputo (1982) and *Journal of Commerce* (1981) both report that plastic resins composed about 6 percent of the typical solid waste stream in 1978, although most estimates are in the 4–5 percent range. [See, for example, Huls and Archer (1981), JRB Associates (1981), and Plastic Bottle Institute (1981b).] Plastics composed 2–3 percent of the solid waste stream in the early 1970s. *NCRR Briefs* (1980) reports that other components in the typical municipal waste stream are by weight 35 percent paper, 16 percent yard waste, 15 percent food waste, 10 percent glass, 9 percent ferrous metals, 3 percent rubber and leather, 2 percent textiles, 1 percent aluminum and other nonferrous metals, and 5 percent miscellaneous materials. The separation of plastics from the municipal waste stream is therefore complicated because plastics compose a relatively small percentage of the total waste stream.

There are three levels of separation of plastics from the municipal waste stream. The first involves the separation of plastics, paper, and other similar weight items from the metallic and heavy portions of the waste stream. In almost all cases, this separation step requires that the entire waste stream be shredded into small pieces after some initial hand sorting.[4] While there are numerous specific methods available for shredding, Leidner (1981) reports that hammermills are used most extensively in the size reduction of municipal solid waste. One problem with the shredding step has been the possibility of explosions from volatile materials entering the shredding equipment. Following the shredding step, the small pieces are separated by air classification, magnetic separation, and screening. Air classification takes advantage of the different densities of waste materials, while magnetic separation recovers the ferrous metals. Screening can be used for separation because different materials vary in particle size following the shredding step. Plastics and paper usually end up in the same fraction following this initial separation step.[5] If plastics are to be recycled in a quaternary and, in some cases, tertiary sense, this level of separation will be sufficient.

The second level involves the separation of plastics from paper. Processes available for this step are divided into dry and wet methods. The dry methods are based on the different reactions of paper and plastics to heat and electrical charges. One method uses a hot rotating cylinder. Thermoplastic materials tend to adhere to the hot cylinder wall and are scraped off by a knife to a conveyor belt. Thermosetting resins will, of course, not adhere to the hot walls, which can be considered an advantage of this process since most secondary recycling operations do not effectively utilize thermosets. A problem with this method is that paper may tend to stick to the melted plastics. Another method uses hot air to shrink the plastics, which are then removed from the paper by screening or air separation. Plastics and paper also tend to hold static electrical charges differently, which can be used to advantage in a drum-type structure that is electrically charged. Wet processes utilize the water absorption differences between plastics and paper.[6] This level of separation will be sufficient for quaternary, tertiary, and many secondary recycling processes.

A third level involves the separation of individual plastic resins. Processes at this level include sink/float methods, processes that utilize the varied wetting characteristics of different resins, solvent extraction, and electrostatic methods. Sink/float methods depend on the different densities of resins.

Different wetting characteristics of resins allow separation by flotation. Solvent extraction is based on the property that different resins when dissolved in a solvent will not usually mix together, but rather will form different phases, which facilitates separation. Electrostatic methods depend on two characteristics of resins. First, as temperature increases, the ability of plastics to accept and hold a static charge decreases. And second, the temperatures reached by different plastics vary with electrical properties.[7] More technologically advanced separation methods that are currently being explored include cryogenic grinding with magnetic separation and separation assisted by the use of microorganisms.[8]

While several separation technologies currently exist and other more advanced processes are in the development stage, most reports do not hold high hopes for the separation of plastics from the municipal waste stream. The general opinion among the technical experts is that once plastic wastes enter the municipal waste stream, the separation of plastics from other municipal wastes is difficult both technically and economically, and the separation of specific resins is not now, nor is it expected to be in the reasonable future, economically viable.

PRIMARY RECYCLING

Primary recycling of plastic wastes, which converts the waste into products similar in characteristics to the original product, has been practiced by the plastics industry for some time. Leidner (1981) estimates that the industrial generation of plastic wastes was about 3 million tons in 1980, much of which was recycled by the plastics manufacturer or by reprocessors. Leidner also estimates that about 4.7 percent of total plastics production is recycled by reprocessors and over 8 percent recycled in manufacturing plants. Approximately 4 percent of the output of resin producers becomes a recyclable waste, 2.4 percent for fabricators, and 5.0 percent for plastic converters.[9]

However, it is generally agreed in the literature that, because of constraints on waste contamination and as yet inadequate separation processes, primary recycling is feasible only for relatively clean industrial scrap plastics. No processes or cost estimates were found in the literature for the primary recycling of the more contaminated manufacturing waste, usually referred to as manufacturing nuisance plastics, or (with one exception) for any postconsumer plastics.[10] Therefore, primary recycling can at the present time be ruled out as a viable technological or economic solution to postconsumer and manufacturing nuisance plastics. Given that this analysis is concerned with postconsumer and manufacturing waste that is currently being disposed of, primary recycling will not be considered further in this book.

SECONDARY RECYCLING

Secondary recycling, or recycling plastic products into new products with less demanding physical and chemical characteristics, has received significant attention in recent years. Countries other than the United States have been in the forefront of the development process, especially Japan and Western European countries where the relative costs of disposal and raw materials are greater than in the United States.

There are numerous sources that give excellent discussions of the technical issues to be considered in secondary recycling.[11] This section reviews the major conclusions of those studies by summarizing the approaches that have been used in secondary recycling, giving a brief description of some of the more well known machinery used, and discussing some of the major areas where secondary recycling markets have been developed or have an excellent potential for development.

Approaches to Secondary Recycling

Plastic wastes that are potentially eligible for secondary recycling come from four different sources: a) postconsumer plastics that have been diverted from the municipal waste stream because of, for example, deposit laws or community recycling collection centers; b) postconsumer plastics that are produced as residue in the recycling of products, for example, residue from the shredding and recycling of automobiles; c) manufacturing wastes containing single resins; and d) manufacturing wastes containing a mixture of resins. The level of contamination associated with each waste type will vary, which will limit the types of secondary recycling that can be applied.

Leidner (1981) cites three major technical problems that face secondary recycling. First, the waste streams that are being considered for secondary recycling usually are contaminated with materials such as dirt and metals, which can do significant damage to equipment that is normally used in plastics manufacture. And as discussed in an earlier section, separation technologies that could potentially remove the contaminants are not considered economically viable. Second, in many cases different plastic resins are incompatible in mixtures, which results in poor mechanical properties. Third, the characteristics of the available waste feedstocks are not usually uniform over time. The processes must therefore have the flexibility to process varying waste mixtures.

Four different approaches have been developed to overcome these technical obstacles. First, equipment that is normally used in the processing of virgin thermoplastic materials has been altered and used to heat and reform thermoplastics as a segregated waste stream or as part of a mixture with thermosetting resins and other materials. These alterations prevent the equipment from being damaged by the nonplastic materials in the waste stream. Conceptually, the processes are relatively simple in that they melt the thermoplastics in the waste mixture by heat and/or pressure and then discharge the melt into a mold for cooling. One of the major problems with this form of recycling is, however, the relatively low mechanical properties of this type of plastics mixture. To help overcome this problem, these secondary products are usually bulky, like fence posts, heavy drainage gutters, and other items that might substitute for wood or concrete.

Second, the waste may be chemically altered to overcome the problems posed by the incompatibility of different resins in mixtures. Two approaches have been attempted. In one, compatibilizers have been used to overcome the natural tendencies of different resins to form phases when mixed. The other method involves the cross-linking of the different molecular structures. In this process the thermoplastic resins are actually converted to thermosets, which, of course, increases their properties under higher temperatures, but also increases the difficulty of their future recycling.

Third, the waste thermoplastics can be melted and co-extruded or co-injected into moldings with virgin resins. In these processes, virgin resins are injected first and are forced to the perimeter of the mold. The recycled plastics with inferior physical and chemical properties are then injected in the center of the mold where the inferior properties are not usually of crucial importance.

The fourth process uses finely ground waste plastics as fillers in mixtures with virgin resins. This process is the only secondary process that is applicable to thermosets. The thermosets, which act as fillers, are not particuarly different from other inexpensive nonplastic filler materials and therefore do not offer significant recycling benefits. Mechanical properties of the mixture are reduced as the percentage of filler material increases.

Examples of Secondary Recycling Processes

All of these secondary approaches, with the exception of chemical modification, have at some time been used commercially. This subsection reviews some of the technical characteristics of some of the better known technologies.

The Mitsubishi "Reverzer"

The Mitsubishi Reverzer developed in Japan is one of the best known secondary recycling processes.[12] This process heats and reforms mixed plastic wastes into products such as fence posts, large cable reels, and other bulky products usually made from wood. Drain pipes and gutters, floor tiles, and other large products that benefit from the nonbiodegradability of plastics can also be manufactured with the Reverzer. The products can be sawed, nailed, and handled in much the same way as wood products, but are much more resistant to moisture and therefore for some applications can be considered superior to wood.

The major advantage of the process is that it can accept up to 50 percent non-thermoplastic materials, including glass, sand, and metal fragments. Further, the process is quite resilient to variations in the types of thermoplastic resins used. According to *Materials Engineering* (1978) the process also includes a patented melting process that avoids the production of chlorine and cyanide gases, which can be a major environmental problem with the melting and processing of polyvinyl chloride (PVC) and polyurethane materials.

The Regal Converter

The Regal Converter, which was developed in the United Kingdom, can also accept up to 50 percent non-thermoplastic materials. In this process the waste material is chopped into small pieces and is then heated and rolled into sheets. The recycled product can be used in much the same way as wood particle board. One of the major benefits of the Regal Converter is that the different waste materials do not have to be mixed or homogenized. A derivative of the Regal process — the Kabor K-Board — was operational in the United Kingdom in the early 1970s, but was closed in 1974 because the sheet product was no longer competitive.[13]

Other processes similar to the Regal Converter and the Mitsubishi Reverzer include the Klobbie (developed in the Netherlands), the FN Machine (developed in France), the Flita System (developed in West Germany), the Remaker (also developed in West Germany), and a process by Japan Steel Works Limited. [See Leidner (1981) for details.]

Alternative Board Producing Processes

Other processes are being developed to recycle scrap plastic materials from automobile manufacture and recycling. For example, the Upjohn Company has been involved in the development of processes to make particle board from plastic auto scrap. [See, for example, McClellan (1983).] Another example of this work is the current effort by Oak Ridge National Laboratory (ORNL), the Plastics Institute of America, and various universities to use the residue from automobile shredder operations to make particle board. This work is being funded by the Energy Conversion and Utilization Technologies (ECUT) Program within the U.S. Department of Energy. [See Hawkins (1982), Plastics Institute of America (draft), and *Machine Design* (1984) for details.] The potential for the recycling of automobile shredder residue is examined in detail in Chapter 6 of this book.

Goodyear PET Recycling Equipment

The Goodyear Tire and Rubber Company, a major producer of polyethylene terephthalate (PET), has developed a process for the grinding and cleansing of PET beverage bottles [see Goodyear Tire and Rubber Company (undated) for details]. The process is mainly aimed at the removal of paper, metal, and other non-PET plastic components of the typical beverage bottle. Products produced from PET recycling are varied and include fiberfill for clothing, sailboat parts, and assorted household products such as sinks, countertops, and tubs (see Bennett, 1985). Similar technologies have been, or are being, developed to process PET bottles. Chapter 6 of this book discusses the recycling of PET beverage bottles in detail.

Thermoset Recycling Processes

Thermosetting resins are usually considered to be nonrecyclable at a secondary level because their cross-linking bonds prevent their being melted and re-formed. Work has been done, however, to grind waste thermoset materials and use those materials as inert fillers with virgin resins.

In the typical thermoset recycling operation the material is ground to a very fine powder. That powder is then mixed with virgin material in a typical molding process. Since the material is ground very finely, the molding step is not significantly altered by the addition of up to 15 percent thermoset waste. With additions of between 15 and 25 percent, the flow of the material will be noticably stiffer. According to Bauer (1976), the addition of up to 15 percent thermoset fillers with most virgin resins results in little change in physical properties, i.e., impact, tensile, and flexural strength. However, impact and flexural strength decrease sharply when the percentage of filler goes above 20 percent. Heat resistance of the mixed material tends to increase, while chemical properties of the mixture are the same as those of the virgin material. The addition of more than 15 percent ground thermosets can produce some slight irregularities in the appearance of the finished material.[14]

Potential Secondary Recycling Markets

There are numerous examples of where secondary recycling is currently used and has a good potential for use. The processes mentioned above that produce products similar to those made from wood currently utilize contaminated manufacturing waste as the predominant raw material. Automobile shredder residue offers a large and potentially exciting material source for secondary recycling. Bottle deposit laws in several states have created a uniform and stable supply of beverage bottles as a waste material source.[15] In addition, high-density polyethylene has been recovered from deposit milk bottles.

Other current markets include the recycling of polypropylene from scrapped automobile batteries (see, for example, *American Metal Market*, 1979) and the recycling of ABS (acrylonitrile-butadiene-styrene) from telephone equipment by Western Electric (see Chapter 6 for details). International Research and Technology Corporation (1973) discusses the possibilities for using waste plastics as a substitute for sand in concrete and for stone in road construction. That publication also reports that Purdue University has studied the possibility of using ground plastic wastes as a mulch or soil conditioner. However, the researchers had some concern about the possible production of dangerous compounds from the use of plastics in soil.[16]

TERTIARY RECYCLING

Tertiary recycling, or the recovery of basic chemicals and fuels from waste polymers, has several advantages over secondary recycling. First, many tertiary technologies can accommodate rather contaminated waste. The processes will, however, produce higher valued products when used with a segregated stream consisting only of plastics. Second, tertiary recycling is relatively nonpolluting compared with the alternative of disposal, especially incineration. This characteristic may allow tertiary recycling centers to locate close to the sources of waste and thus avoid expensive transportation charges.[17] Third, according to Leidner (1981) the volume of incoming waste can be reduced by as much as 90 percent, while being a net producer of energy. These relative advantages of tertiary recycling have spurred numerous technical studies that have examined the potential for applying tertiary processes to the municipal waste stream and to plastics as a segregated waste. This section reviews those studies by summarizing the technical approaches used and discussing some particular applications of those approaches.

Approaches to Tertiary Recycling

Pyrolysis

By far the most popular tertiary process is pyrolysis, which is applicable to both municipal waste and segregated plastics. Much of the research and development on pyrolysis has been done in four countries—Japan, West Germany, Great Britain, and the United States. West Germany has been particularly active.

Pyrolysis is a process whereby organic materials are heated in the absence or near absence of oxygen in a controlled combustion chamber. The process drives off the volatile components of the waste, while leaving a substance consisting mainly of carbon and ash. Products of the process include combustible gases, which may be used as a chemical feedstock or purified to pipeline quality; liquid products, which can be used as a low-sulfur fuel oil; and residue, which can be used as a fuel or, according to Sperber and Rosen (1974), used as a filtering medium in tertiary sewage treatment or as a particulate component of concrete. The composition and the properties of the products produced will depend on the particular raw materials, the process, and the reaction time and temperature.

In the case of the pyrolysis or municipal waste, Leidner (1981) reports that the typical yield of products will vary according to the process temperature, ranging between 500 and 900°C. At 700°C, raw municipal waste that contains about 1.5 percent plastics has been shown to yield 11.5 percent residue, 23.7 percent gas, 1.2 percent tar, 0.9 percent light oil in gas, 0.03 percent free ammonia, and 55.0 percent liquor. Leidner (1981) further reports that since one ton of typical municipal waste requires about 2 million Btus for pyrolysis and in turn produces 8 million Btus of energy products, the process can be self-sustaining. The major commercial product from the pyrolysis of municipal waste is low-Btu fuel gas. Problems with the process include varied product quality due to significant variations in incoming municipal wastes and potential environmental problems posed by the production and disposal of liquor, which contains 94–100 percent water.[18]

The products from pyrolysis are of higher value if the waste stream consists only of one resin or a mixture of plastic resins. Products from the pyrolysis of plastics can be used as gaseous or liquid fuels, or basic feedstocks to the chemical industry. Significant work has been done on the pyrolysis of polystyrene, PVC, and, in particular, polypropylene. Polystyrene and PVC can be completely pyrolyzed at temperatures between 400 and 500°C. The production of gases from the processes increases with temperature. Problems with these processes include a buildup of carbon residue on some process parts and variations in product specifications that result from variations in the quality of input materials.

Hydrolysis

An alternative tertiary process is hydrolysis in which plastic wastes are decomposed by chemical rather than thermal means. According to Leidner (1981), hydrolysis has several advantages over pyrolysis, including a more uniform product, fewer requirements for the separation and purification of products, and potentially smaller capacity requirements and operational size. The main disadvantage of hydrolysis is that relatively uncontaminated waste is required. Mixed plastic waste is not usually suitable. Therefore, hydrolysis is mainly limited to the recycling of manufacturing nuisance plastics of a single resin type. However, Leidner and others envision hydrolysis possibilities for scrap polyurethane foam from automobile shredder operations, which usually contains a significant level of impurities, including other resins, dirt, and so on.

Hydrolysis is the reverse action of condensation. When the usually hydrolytically stable plastic resins are subjected to superheated steam for several minutes, the chemical structure of the polymers breaks down to give basic chemicals, which can be used to produce new resins. According to Leidner (1981), the main interest for hydrolysis is in the recycling of waste polyurethanes and thermoplastic polyesters from industrial sources.[19]

Applications

Several commercial examples of pyrolysis processes are available. However, according to Alvarez (1985) none of the four facilities designed to produce energy from municipal waste by pyrolysis are operational. Those facilities were, or are, in Disneyworld, Florida; San Diego, California; South Charleston, West Virginia; and Baltimore, Maryland. In all cases the facilities are not being used because of unfavorable economic conditions or have been demolished or replaced with a mass-burn incineration facility.

Significant interest has been generated by a process to pyrolyize segregated plastic waste. The process was developed by Procedyne Corporation in conjunction with the United States Department of Energy. According to Bhatia and Rossi (1982), the process has been commercially adopted by U.S.S. Chemicals in their La Porte, Texas plant, which produces polypropylene. About 94 percent of the available fuel value in the waste polypropylene is converted into usable energy. Liquids account for 90 percent of the fuel, 4 percent is gas, and the remaining 6 percent is used to heat the pyrolysis reactor. The liquid fuels substitute for No.6 and No.2 fuel oil.[20]

Leidner (1981) discusses several proprietary pyrolysis processes for segregated plastic wastes. Those processes use a variety of resin feedstocks, including polyethylene, polypropylene, polystyrene, PVC, and other polyolefins. Products include heavy oil, waxes, fuel oil, and naphtha kerosene fuel. Most of the processes reviewed were developed in either Japan or West Germany. A detailed analysis of the Mitsubishi Thermal Decomposition Process showed that for that process the production of oils far exceeds the production of gases for all resins input. For example, at one extreme, polypropylene is processed into 1.8 percent gas and 97.1 percent oil, while, at the other extreme, a mixture of several resins including polyethylene, polypropylene, and polystyrene is processed into 11.9 percent gas and 73.2 percent oil.

An example of a commercial hydrolysis plant is General Motor's operation to recycle polyurethane foam. The foam is hydrolyzed with steam at about 316°C to produce polyols. The products can be reused following cooling and filtering [see Leidner (1981, pp. 269–70) for details.]

QUATERNARY RECYCLING

Currently, the most popular approach to recycling plastics and other combustible solid wastes is quaternary recycling, or the retrieval of the energy content of the waste by burning. While burning waste to reduce its volume is certainly not new, the idea of retrieving energy from waste is a relatively new concept, having been adopted on a large scale by the Europeans in the early 1960s. The main problems with older incinerators, which did not retrieve heat energy, were their inabilities to reach high enough temperatures to burn municipal waste effectively and their production of large quantities of particulate emissions. Modern technology has made great strides toward correcting these problems and has therefore made incineration with heat recovery a viable recycling process for both segregated plastic wastes and plastic wastes as part of the municipal waste stream. Municipal waste incinerators can reach temperatures in the 900–1,000°C range with forced air injection. And modern advances, such as scrubbers and electrostatic precipitators, have drastically reduced the level of air pollutants from incineration equipment. Much of the research on, and the application of, modern incineration methods has taken place in Western Europe and Japan. The United States is now, however, in a catch-up role as the potential benefits of incineration with heat recovery make those processes increasingly viable from technological and economic perspectives.

This section reviews the technological approaches that have been used to retrieve heat from solid wastes and discusses some current examples of those approaches. Incineration with heat recovery is discussed in two parts: the incineration of municipal solid waste, and the incineration of plastics as a segregated waste stream.[21]

Incineration of Municipal Solid Waste

The incineration of municipal solid waste with heat recovery has the advantage of recovering a portion of the heat value in the waste, while reducing the incoming waste by about 80 percent in weight and about 90 percent in volume. The remaining residue can then be landfilled.[22]

When compared with conventionally used fuels, the Btu content of the typical municipal solid waste is low. Leidner (1981) estimates that the typical municipal waste contains between 4,000 and 6,000 Btus per pound in its raw, unprocessed state. Diaz, Savage, and Golueke (1982) put the estimate at 4,500 Btus per pound. Bituminous coal, by comparison, contains about 13,100 Btus per pound, distillate fuel oil about 19,160, and residual fuel oil about 18,990. However, when municipal waste is compared with unconventional fuels, such as peat at 3,586 and oil shale at 6,300 Btus per pound, municipal waste is competitive in heating value. Table 2.1 gives the heating values of various

TABLE 2.1. Btu Values of Different Fuel Types

Material	Heating Value Btus/lb	Source
Unprocessed municipal waste	4,000 to 6,000	1
Unprocessed municipal waste	4,500	2
Bituminous coal	13,100	3
Anthracite coal	12,700	3
Distillate fuel oil	19,160	3
Residual fuel oil	18,990	3
Woodflour	8,520	1
Peat	3,586	4
Oil shale	6,300	4
Sewage sludge	7,500	4
Paper	7,590	4
Polyethylene	20,050	4
Polystyrene	17,870	4
Polyvinyl chloride	7,720	4
Polycarbonate	13,310	4
Unsaturated polyester	12,810	4
Epoxies	14,430	4
Polypropylene	20,030	1
Melamine-formaldehyde	8,310	4
Urea-formaldehyde	7,680	4
Polyvinyl fluoride	9,180	4
Polyvinyl alcohol	10,760	4
Polyurethanes	10,180	4

Sources: 1, Leidner (1981, pages 298–299); 2, Diaz, Savage, and Golueke (1982, Volume II, page 11); 3, American Petroleum Institute (1984, Section XV); 4, Thorne and Griskey (1972, pages 98–99).

conventional and unconventional fuels. Note that most common plastic resins contain heating values in the range of 7,000–20,000 Btus per pound, with about 12,000 Btus per pound being the average.

Two major types of incineration equipment have been used in the United States to recover heat from municipal solid waste: waterwall incinerators, and modular incinerators. Each broad class of technology has its own good and bad points. However, from the perspective of plastics recycling, the overriding good point is that neither technology requires that plastics be separated from other combustible municipal wastes.

Waterwall Incinerators

There are three basic types of waterwall incinerators. One burns raw refuse and does not require any preparation of the waste. A second type burns all of the municipal waste stream after a preliminary shredding operation. A third type burns only refuse-derived fuel (RDF). RDF refers to the combustible portion of the municipal waste stream and is obtained by first shredding the entire stream, subjecting the stream to magnetic separation to remove ferrous metals, and separating the light and heavy remaining fraction by air separation. The light fraction is sold as a RDF and can be substituted for, or used in conjunction with, conventional fuels such as coal in conventional boilers. However, the use of RDF in conventional boilers presents problems because of relatively high moisture and ash content, relatively low heating value, and contamination with fuels that may pose pollution and operational problems. According to Diaz, Savage, and Golueke (1982), the heat content of RDF varies between about 5,200 and 6,700 Btus per pound, depending on the degree to which the RDF is screened from other municipal waste materials.

Waterwall incinerators are given their name because their walls consist of closely spaced tubes through which water circulates to recover heat from waste combustion. In addition, the incinerators are designed to retrieve the heat from exhaust gases through conventional waste heat boilers. The recovered heat can be used for a variety of uses, including the generation of steam, hot water, hot air, or fed into turbines for electricity generation.

Diaz, Savage, and Golueke (1982) identify three major problems with waterwall incinerators. First, the systems are usually only available about 80 percent of the time because the grates on which the combustion occurs must be cleaned quite often. This occurs because of the significant quantities of non-combustible materials in the waste stream. Second, the heat-transfer surfaces require frequent cleaning because of the accumulation of slag and ash. Third, corrosion has been a significant problem. One of the main contributors to the corrosion problem is the presence of plastic resins such as PVC. The gases produced from the combustion of some resins, when mixed with water, can form strong corrosive acids.

Modular Incinerators

A more recent technological introduction in incineration equipment has been the modular incinerator, which has several advantages over the waterwall incinerator in certain situations. These facilities usually have two combustion chambers to facilitate combustion performance. Most units retrieve heat by conventional waste heat boilers, which use hot combustion gases to produce steam or hot water.

Advantages of the modular incinerator include the following. First, they are factory built, relatively small, and can be transported on highways. Second, the units can be arranged in clusters making expansion of a facility relatively simple. These characteristics make modular incinerators attractive to small communities that do not need the larger waterwall facilities. They are also advantageous to operations that are not ready to make a long-term and capital-intensive commitment to the incineration of their waste materials.

Ducey et al. (1985) discuss twenty common problems in small (under 50 tons per day) waste-to-energy plants. (Most modular incinerators are in the 25–50 tons per day range.) The most common problem was refractory damage due to excessive heat. In some states with stringent emission regulations, air pollution was a problem. Some facilities had difficulties obtaining sufficient quantities of waste materials. Others cited that demand for their steam was lower than expected or fluctuated significantly between seasons. Other problems dealt with specific technical complications.

Diaz, Savage, and Golueke (1982) report that the energy conversion ratios for both modular and waterwall incinerators are about the same, 56 percent for waterwall incinerators and 57 percent for modular incinerators. Facilities that use RDF can expect conversion ratios of about 57 percent.

Current Facilities

Waste Age (1984) and Alvarez (1985) give information about the current facilities that burn municipal solid waste for heat recovery. The 1984 publication also lists other types of municipal resource recovery facilities. The 1985 publication focuses on RDF, pyrolysis, mass-burning waterwall furnaces, and other incineration with heat recovery. The survey reported in Alvarez (1985) includes facilities that are operational, in the shakedown phase, under construction, or under contract.

The survey reports a total of 34 facilities producing RDF with a total capacity of 40,230 tons per day. The facilities range in size from 100 to 3,000 tons per day and are located in 19 different states. New York state has the largest total RDF capacity at 7,130 tons per day at five different facilities. There are 27 waterwall facilities with total capacity of 27,150 tons per day. Facilities range in size from 120 to 3,000 tons per day and are located in 16 different states. There are 12 conventional incinerators that have waste heat recovery systems and have a total capacity of 4,248 tons per day. Those facilities range in size from 48 to 1,000 tons per day and are located in 9 different states. Finally, there are 31 modular incinerator facilities with a total capacity of 3,083 tons per day. Those facilities range in size from 7 to 360 tons per day and are located in 21 different states. Paul (1985) reports that the total number of U.S. waste-to-energy plants operational, under construction, or in the planning stage in 1985 increased by 36 percent over the 1984 level.

According to the Environmental Protection Agency, residential and commercial solid waste totaled 138.9 million tons in 1978 or about 381 thousand tons per day.[23] Further, Diaz, Savage, and Golueke (1982) report that municipal waste constitutes only about 53 percent of the total urban waste stream. Given that the total capacity for quaternary resource recovery of all types will be only about 75 thousand tons per day when all capacity comes on line, there is ample room for expansion in the quaternary resource recovery area.

Incineration of Segregated Plastic Wastes

Approaches and Problems

As indicated in Table 2.1, plastic wastes have significantly higher heating values than does municipal waste. On average, plastics are about equivalent to coal. Therefore, in those cases where plastics are collected outside of the municipal waste stream, an attractive recycling method may be burning with heat recovery.

However, a number of problems plague this form of recycling. Standard incinerators, which are designed to burn municipal waste, cannot be used to burn segregated plastics. The combustion of plastics requires three to five times the amount of oxygen that conventional incineration requires. The absence of sufficient air will result in the excess production of soot, which has obvious environmental implications and can seriously damage the equipment. Moreover, conventional incinerators cannot handle the excessive heat produced by plastics combustion. Finally, precautions must be taken to counter the production of toxic fumes that results from the burning of some resins.

Available Processes

Several incineration processes have been developed that overcome these technical obstacles. According to Leidner (1981), Mitsubishi Heavy Industries, Ltd. has developed a continuous-rotary-kiln type incinerator that is suitable for plastics. Another process has been developed by Takuma Boiler Manufacturing Company. In the United States, Industronics Incorporated has developed a process to burn plastics that is currently being used to retrieve the energy content from PET beverage bottles. [See Industronics Incorporated (1982) for details.] The burning of PET bottles has also been tried by Crown Zellerbach Corporation in Vancouver, Washington. In this example, PET is used as a supplement with low-quality wood fiber to provide steam to power equipment.[24]

DISPOSAL

Plastic wastes that do not enter one of the above discussed recycling streams are inevitably disposed of by incineration and/or landfill.[25] The technological concerns about disposal are mainly environmental and are discussed in the following section. This section presents a brief discussion of the processes currently used for disposal. Information is also provided on the extent to which each process is currently utilized.

Incineration Without Heat Recovery

Aside from the environmental problems posed by the incineration of waste plastics as part of the municipal waste stream or as a segregated waste, the technological problems faced by incineration facilities with no heat recovery are similar to those faced by systems with heat recovery. The environmental problems are discussed in the following section. See the previous section for the remaining technological concerns.

Landfill

Aside from the potential environmental dangers, the technological problems posed by sanitary landfill are trivial when compared to the problems faced in the recycling or incineration of plastic wastes. In the typical sanitary landfill, waste materials are either spread over a flat area of land or put into a trench that has been dug. The wastes are then compacted to reduce volume. Plastics can be a problem in that they resist compaction relative to other waste materials. The compacted waste is then covered daily with earth in depths of about six inches. Leidner (1981) reports that a minimum of 24 inches of compacted soil is recommended as a final cover for the waste. A key aspect of designing a landfill site is providing for good water drainage to help prevent seepage or erosion of the soil.[26]

Current Waste Facilities

While no detailed information was found to indicate the exact percentage of total U.S. solid waste that is currently disposed by sanitary landfill, it is general knowledge that landfill is the predominant disposal method. *Waste Age* (1985b) reports that most landfill operations are fairly small, usually under 250 tons per day.[27] In addition, Savino and Gould (1984) give data on how the nation's 16 largest cities manage solid waste. According to the survey reported in that article, ten of the cities use landfill exclusively for disposal of their solid waste. Of the six remaining cities, two dispose of 50 percent of their waste by incineration and the remaining cities employ incineration to a lesser extent — between 9 and 34 percent. Chicago is the only top 16 city that uses resource recovery, which is applied to about 25 percent of their total solid waste.

ENVIRONMENTAL IMPACTS

One of the major arguments that has been used to support the recycling of plastics concerns the environmental impacts of plastics in incineration and landfill operations. And there is evidence to suggest that the general population perceives plastic materials to be very harmful to the environment. Opinion Research Corporation (1975) found from an interview of 192 individuals that 65 percent of the interviewees believed plastics to be most harmful to the environment of all commonly used materials. Paper and steel were each selected by 7 percent as the most harmful, while aluminum was in fourth place with 4 percent. Wood was believed to be least harmful to the environment. The characteristics of plastics cited as being most damaging were toxic fumes given off during incineration and during the manufacturing process and the nonbiodegradability of most resins. The interviewees included solid waste officials, fire chiefs, news people, energy and environmental officials, health officials and educators, environmental educators, and community leaders.

This section discusses the different technical views about the environmental impacts of plastics in both disposal and recycling operations. As will become clearer, the environmental impacts of plastics when disposed or recycled are not as obvious as many individuals may believe.

Secondary Recycling

The environmental impacts of the secondary recycling of plastic wastes are obviously minimal. In most secondary operations waste thermoplastics are melted and reformed by themselves or in a mixture with other materials. While there is the possibility that some toxic fumes may be produced in the melting operation, this risk is minimal when compared to the potential environmental damages caused by other disposal and recycling alternatives.

However, secondary recycling is only a postponement and not a solution to the environmental problems posed by plastic wastes. It is generally agreed in the technical literature that the secondary recycling of thermoplastics will to some extent damage the chemical bonds of the resins. This degradation will ultimately force the recycled resins to enter a tertiary, quaternary, or disposal stream.

Secondary recycling will reduce the long-term environmental problems caused by plastics disposal only to the extent that the secondary products replace products that would have been produced with virgin resins. A simple example, illustrated in Figure 2.3, may help to explain this concept. Suppose we have three products, X, Y, and Z. For simplicity, assume that total demand for each of these products is constant over time at 100 units. Each year consumers discard 10 units of each product, implying that total production of each product is 10 units per year. Further, assume that product X can be produced only from virgin resins. Product Y can be produced from either virgin resins or recycled resins from the discarded product X. Product Z is normally produced from wood, but can also be produced from recycled resins from product X. In each case one unit of resin is required for the production of each unit of the product.

In Case A of our example, one half of the discarded units of product X are recycled in a secondary sense to produce product Y. The implication of this recycling is that the demand for virgin resin in the production of product Y is 5 units, or 5 units less than if no recycling occurred. In addition, the disposal or tertiary or quaternary recycling of resins from both products X and Y is equal to 15 units. Note that in the absence of secondary recycling, the flow to disposal and tertiary and quaternary recycling would equal 20 units.

In Case B of our example, one half of the discarded units of product X are recycled in a secondary sense to produce product Z rather than product Y. The implication of this change is that 20 units of resin production are now required to satisfy the production of products X and Y, rather than the 15 units in Case A. Further, note that the flow of plastic wastes entering tertiary or quaternary recycling or the disposal stream now equals 20 units instead of the 15 units in Case A. Therefore, the secondary recycling of waste plastics to produce a product that is normally produced from a material other than plastic resins will have no impact on either the total production or disposal of plastic resins.[28]

Variations of this example can, of course, be formulated in which the lifetimes of the different products differ, thus altering the flows of plastics to disposal or tertiary or quaternary streams in the short term. For example, the reuse of plastic packaging materials, which usually have life times of less than a year, into products that substitute for wood, which may have life spans of 25 years, will obviously divert those wastes from further recycling or disposal for that 25 year period. However, in the long term the flows will stabilize and the conclusions presented in our simple example will hold. Therefore, the benefits of secondary recycling in eliminating the potential environmental damages caused by disposal or by tertiary or quaternary recycling are nil in the long term if the secondary products do not replace products made from virgin resins. Further, and unfortunately, secondary recycling will seldom produce products that compete with products produced from virgin resins. This, of course, does not negate the short-term environmental benefits provided by diverting disposal or tertiary or quaternary re-

FIGURE 2.3.   The Impact of Secondary Recycling on Plastic Waste Production

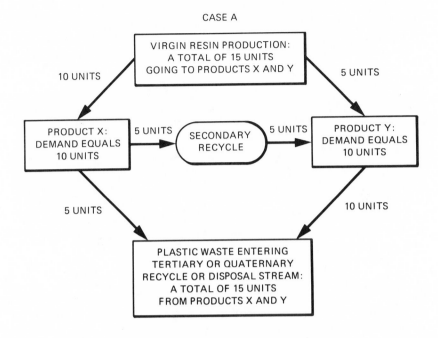

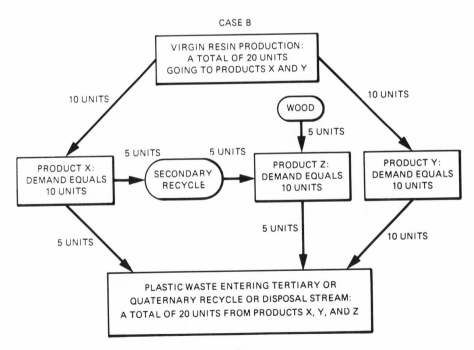

cycling to future time periods. The value of this diversion will, of course, depend on the assumed discount rate. Neither does it negate the nonenvironmental benefits provided by secondary recycling.

Tertiary Recycling

In the tertiary recycling of plastic wastes, in which fuels and basic chemicals are produced, the environmental problems may be less severe than incineration or landfill, but will not be nil. Leidner (1981) states that pyrolysis is a contained process and therefore does not cause air pollution. However, according to Leidner, the liquor produced in the pyrolysis of municipal refuse may present a significant waste disposal problem. Further, from a broader perspective, one must examine the environmental implications of the use of the products produced from the pyrolysis process—i.e., basic chemicals and fuels. When this broader perspective is taken, it is not clear what the environmental implications of tertiary recycling may be.

Incineration

Significant concerns about air pollution have been raised in regard to the incineration of plastic wastes as a segregated waste or as a component of the municipal solid waste stream. This is true for both incineration with and without heat recovery. [According to Vaughan et al. (1975), the addition of a boiler to an incinerator will not have a significant effect on the emission factors.]

Consider plastics in the municipal waste stream first. In this case, there are characteristics of plastics that make them both suitable and unsuitable for incineration. On the one hand, plastic resins contain little or no sulfur, which can produce sulfur dioxide when burned, and there is little ash residue remaining after incineration. Further, plastics contain higher Btu contents than any other common waste entering the waste stream. The mixture of plastics with other waste materials therefore helps to elevate the heat of the process and thus facilitates the combustion of other materials, especially those containing a high moisture content. If plastics were not part of the municipal waste stream, it is likely that the residue from the incineration of the remaining wastes would be increased, thus increasing the volume of residue requiring landfill.

An often quoted 1971 study by Kaiser and Carotti examined the impacts of 2 and 4 percent additions of several common resins to the typical municipal waste stream. The study utilized a 220 ton-per-day municipal incinerator and concluded that a conventional municipal incinerator will perform satisfactorily with normal refuse containing up to 6 percent plastics. At these percentages, no significant increase in smoke occurred, indicating that a conventional furnace can provide sufficient quantities of air for combustion. The odor from flue gases was not increased by the burning plastics, with the exception of hydrogen chloride produced from polyvinyl chloride. The test was done with four commonly used resins: polyethylene, polystyrene, polyurethane, and polyvinyl chloride. In a later study by the same researchers at the same facility, the test was repeated with PET. That study reached the same basic conclusions.

On the other hand, the combustion of some plastics, especially PVC, can produce significant levels of hydrogen chloride. About one half of the weight of PVC is composed of hydrogen and chloride, which combine during burning to form hydrogen chloride. The hydrogen chloride then reacts with water to form hydrochloric acid. The combustion of plastics can produce other potentially toxic products, such as nitric oxides and a host of pollutants caused by the incineration of plastic additives. Further, in a more recent study reported in Velzy (1985), there is some concern in that investigators have found dioxins in precipitator fly ash at some energy-from-waste plants. According to the article, it has been suggested that some dioxins " . . . are formed in direct relation to the combustion of PVC plastics contained in municipal solid waste" (Velzy, 1985, p. 188). The article goes on to state, however, that this assumption about PVCs is most probably incorrect.

Most experts conclude that plastics do not pose a significant environmental hazard when burned with other municipal wastes. While the experts acknowledge that harmful air pollutants can be produced, they argue that a properly designed and operating incinerator equipped with scrubbers and other available stack gas controls will be able to burn plastics while maintaining effluents within air pollution standards. Others argue, however, that while plastics can be burned safely, many incinerators are poorly designed and operated, which worsens the problem.[29] Still others argue that even if some pollutants are produced by the incineration of plastics, their levels will be insignificant compared to other major sources of air pollution.[30]

The problems found in the incineration of plastics with other municipal wastes will be exacerbated when burned as a relatively pure plastic waste. Leidner (1981) reviews several problems associated with the incineration of predominantly plastic wastes. As discussed above, there will be problems with the production of toxic fumes depending on the specific resin burned. If sufficient air is not provided to the incinerator, soot will be produced. There may be problems in disposing of water that has been acidified by hydrogen chloride. Finally, lead and cadmium salts, which are sometimes used as stabilizers in PVC, will remain in the residue and cause disposal problems.

Plastics in Landfill

The experts differ sharply in their views about the environmental impacts of plastics in landfills. On the one hand, some experts view plastics as being harmless. Plastics are not soluble in water, they decompose slowly (over a period of 10 to 30 years according to Vaughan et al., 1975), and the decomposition products are usually inert. In addition, some have argued that plastics provide structural support for the landfill because they do not degrade rapidly. Biodegradable landfill materials, which are reduced in volume as they decompose, can result in landfill sinking. The structural support provided by plastic wastes can help alleviate this problem and make the landfill site more functional once the landfill is closed.

On the other hand, some experts have argued that plastics do pose several environmental problems in landfills. First, plastics are more bulky on a weight basis than most other landfill materials and thus require more space per pound. Second, while plastics decompose slowly and in general do not give off hazardous or distasteful odors or gases, decomposition does occur, which can potentially result in air and water pollution. According to Leidner (1981), decomposition of plastics results from oxidation, hydrolysis, microorganism attack, and stress cracking. Different resins have different susceptibilities to decomposition. And the elevated temperatures in the landfill due to the decomposition of the other materials can result in faster decomposition of plastics by oxidation and hydrolysis. Third, the same physical characteristic that helps give structural support to landfills can also be harmful. If the percentage of plastics is too high or if plastics are not distributed well, the landfill may develop spongy areas. Fourth, plastics, because they are difficult to compact, will tend to expose other degradable wastes to water. Vaughan, Anastas, and Krause (1974) estimate that the presence of plastics will allow 5 percent more water to contact other materials in the landfill due to infiltration and percolation. Water pollution from the landfill is thus more difficult to manage.[31]

CONCLUSIONS

The primary purpose of this chapter has been to present an overview of the numerous technological issues that influence the decision to dispose of or recycle plastic wastes. This overview is important not only because it gives the reader an appreciation of the difficult technological barriers that have been, and must be, overcome, but also helps define the relevant economic and institutional parameters to be considered in the remainder of this book.

Three main focuses have been considered. The first examined how the physical and chemical properties of different types of plastic wastes affect the recyclability of those wastes. The overriding conclusion is that plastic waste cannot be considered homogeneous, but must be defined by its particular resin content and the degree to which the waste is contaminated with other materials. Resins differ widely in their physical and chemical properties, which affects their applicability for different recycling processes. Further, our current abilities to separate plastics from other waste materials and to separate particular resins limit the recycling alternatives for certain waste streams. It is therefore important that information about future levels of plastic wastes be available by resin type and by level of contamination. This is the topic of Chapter 4 of this book.

The second focus was on the technical characteristics and requirements of different recycling and disposal alternatives. While primary recycling can be ruled out for almost all postconsumer and manufacturing nuisance plastics, there are numerous secondary, tertiary, and quaternary recycling processes currently available, each having various advantages and disadvantages. The technical applicability of these numerous processes to recycle plastics depends on the particular waste stream. And while this large variety of available processes increases the chances that some form of recycling will be technically and economically viable, the variety also increases the complexity of the decision process concerning recycling. The numerous economic and institutional complications raised by these and other factors are addressed in the following chapter of this book.

Finally, this chapter addressed the environmental consequences of disposing plastic wastes by incineration and landfill, as well as by recycling. The overriding conclusion is that the technical experts differ on this issue. Many individuals perceive the disposal of plastics to be harmful to the environment and there is evidence to suggest that plastics may pose serious environmental problems in both incineration and landfill. However, the majority opinion is that plastics in the solid waste stream do not pose environmental problems that cannot be effectively managed with available technologies. Another major conclusion is that the potential environmental problems resulting from plastics disposal cannot be completely negated by recycling. Tertiary and quaternary recycling will produce their own associated pollutants. Furthermore, in the case of most secondary recycling processes, waste resins are only temporarily diverted from a disposal stream or a tertiary or quaternary process. It is generally agreed that the chemical structure of most plastics will be degraded by secondary recycling. Therefore, repeated secondary recycling will not be possible. The present technical uncertainty about the environmental impacts of plastics recycling and disposal is unfortunate, because, as is discussed in the following chapter, those impacts will be a major consideration in the public sector's decision to promote plastics recycling.

NOTES

1. See Chapter 4 for estimates of the production of plastic wastes by manufacturers and consumers. The percentage of the manufacturer throughout that becomes a nuisance plastic varies by the stage of production.

2. For a more technical discussion of the differences between thermoplastics and thermosets, see, for example, Leidner (1981, Chapter 1).

3. There are several sources that provide summaries of the separation methods currently available and the various problems associated with those methods. This section draws heavily from the following reports: Huls and Archer (1981), Leidner (1978, 1981), Technomic Consultants (1981), Sperber and Rosen (1974), International Research and Technology Corporation (1973), and Milgrom (1972). See Leidner (1981) in particular. Other works are cited below that focus on particular aspects of the separation problem.

4. See Wilson and Senturia (1975) for alternatives to the shredding step.

5. For more information on this initial separation step see, for example, Vesilind and Pierce (1983), Diaz, Savage, and Golueke (1982), Savage, Diaz, and Trezek (1981), Sherwin and Nollet (1980), and Wilson and Senturia (1975). Diaz, Savage, and Golueke, Volume 1, Chapters 4 and 5 give a particularly in-depth and complete analysis of the shredding and separation steps involved in solid waste resource recovery.

6. For additional information see, for example, Baum and Parker (1974) and Grubbs and Ivey (1972).

7. The following reports address the separation of mixtures of plastics: Pearse and Hickey (1978), *Environmental Science and Technology* (1976), Nagano (1976), Saitoh, Nagano, and Izumi (1976), and Holman, Stephenson, and Adam (1974).

8. See, for example, Basta (1985).

9. See Huls and Archer (1981), Leidner (1981, Chapters 2 and 4), and Milgrom (1972) for further details.

10. Estimates of the costs of recycling and disposal are given in Chapter 5.

11. See, for example, Drain, Murphy, and Otterburn (1981), Huls and Archer (1981), Leidner (1978 and 1981), Technomic Consultants (1981), Milgrom (1972 and 1979), Buekens (1977), Burlace and Whalley (1977), and International Research and Technology Corporation (1973). Leidner (1981) presents a particulary detailed and complete review. Other reports are cited below that do not give an overview of the issues, but address particular aspects of the problem.

12. For more information on this process, see, in addition to the studies cited in note 11, *Materials Engineering* (1978), *Hydrocarbon Processing* (1975), and *European Plastics News* (1974).

13. For more information on the Regal Converter see, for example, Lock (1978) and *European Plastic News* (1974). The Regal Converter is also summarized in most of the general assessments of secondary recycling processes mentioned earlier.

14. For more information on the recycling of waste thermosetting resins as inert fillers see, for example, *Plastics World* (1982), Morel, Richert, and Martin (1980), and Bauer (1976 and 1977).

15. According to Technomic Consultants (1981), the volume of PET for recycling from collection centers in states without deposits laws is insignificant.

16. The Florida State Department of Environmental Regulation has published a 1982 report for the U.S. National Bureau of Standards that lists numerous secondary recycling companies across the United States. Information is given about the materials processed, the products produced, and addresses and phone numbers of the companies. The report also includes firms that recycle materials other than plastics.

17. While from an environmental perspective tertiary processes may be relatively nonpolluting, there may be serious regulatory impediments to locating tertiary recycling facilities close to the applicable sources of waste. See Chapter 3 for more details.

18. For additional discussion of the pyrolysis of municipal waste see, for example, Huls and Archer (1981), Leidner (1981), Poller (1980), Jones (1978), Kaminsky, Menzel, and Sinn (1976), Huang and Dalton (1975), and Baum and Parker (1974). Leidner (1981), Kaminsky,

Menzel, and Sinn (1976), Huang and Dalton (1975), and Baum and Parker (1974) all give technical descriptions of specific processes that were available at the time of their writing.

19. In addition to Leidner (1981) see, for example, Diaz, Savage, and Golueke (1982, Volume II) and Bell, Huang, and Knox (1974) for additional information on the hydrolysis process.

20. For additional background on the Procedyne process, in addition to Bhatia and Rossi (1982), see Machacek (1983), *Chemical Week* (1982b), Lee (1979), and *Energy User News* (1978).

21. The discussion in this section benefited significantly from the excellent discussions of incineration with heat recovery given in Diaz, Savage, and Golueke (1982, Volume II, Chapters 1 and 2) and Leidner (1981, Chapter 7).

22. Authors differ on the complications of disposing the residue from the incineration of municipal waste. Leidner (1981) states that the residue is inert and can be disposed by landfill. However, Diaz, Savage, and Golueke (1982) report that incineration residue may be subject to hazardous waste regulations, at least in California. If the residue is classified as hazardous, the cost of disposal is significantly higher.

23. See JRB Associates (1981).

24. For additional information on incineration with heat recovery see, for example, Huang and Dalton (1975) and Baum and Parker (1974). Vaughan, Anastas, and Krause (1974) discuss the technical problems that plastics may cause in incineration equipment with and without heat recovery.

25. Diaz, Savage, and Golueke (1982) and the Plastic Bottle Institute (1981b) report that as of the mid 1970s, at least 80 percent of U.S. municipalities used open dumping for solid waste disposal. Some open dumping still occurs. Diaz, Savage, and Golueke believe the percentage remains high and the Plastic Bottle Institute maintained that the 80 percent figure was correct for 1981. The Plastic Bottle Institute also reports that in 1981 about 10 percent of all municipal solid waste went to landfills and an additional 10 percent went to incinerators.

The assessment of the current situation is complicated because many facilities call themselves sanitary landfills, but fail to meet the minimum requirements for that classification. *Waste Age* (1985a) reports that of the total landfill facilities reporting in a recent survey, about one third failed to meet some criteria under Subtitle D of the Resource Conservation and Recovery Act. [See *Waste Age* (1985b) for additional information.] However, the Resource Conservation and Recovery Act of 1976 requires an eventual end to open dumping because of its obvious environmental problems. (See Chapter 3 for more information on the Resource Conservation and Recovery Act, especially Subtitle D.) Open dumping is not considered a viable disposal alternative in the long term and is therefore not considered further in this discussion.

26. For additional information on the disposal of plastics in sanitary landfills see Leidner (1981, Chapter 8). Diaz, Savage, and Golueke (1982) present a more detailed discussion of sanitary landfills, but do not specifically address plastics disposal.

27. *Waste Age* (1985b) also gives information on the number of landfills in most states, the sizes of those operations, the ownership and management of the operations, and environmental monitoring of the sites.

28. Economic incentives that may result from a plastic being recyclable are obviously ignored here. It is quite possible that the potential for secondary recycling may spur the production and use of plastics. It can be argued that in many cases the effective price to the consumer is reduced as the product becomes more recyclable, thus increasing quantity demanded (see the following chapter). This effect will, however, have no bearing on the general point to be made here. In fact, if recyclability results in larger quantities of plastics being produced and consumed, the total environmental problems posed by plastics may become worse.

29. See Jensen, Holman, and Stephenson (1974, p. 222).

30. For additional information about the environmental impacts of the incineration of

plastics with other municipal wastes see the following sources: Brady and Andros (1983), Versar Incorporated (1982), *Journal of Commerce* (1981), Leidner (1981), Vaughan, Ifeadi, Markle, and Krause (1975), Vaughan, Anastas, and Krause (1974), Baum and Parker (1974), Huffman and Keller (1973), and Ingle (1973).

31. For more information on the environmental impacts of plastics in landfill see *Journal of Commerce* (1981), Leidner (1981), the Plastic Bottle Institute (1981b), Vaughan, Ifeadi, Markle, and Krause (1975), Baum and Parker (1974), Vaughan, Anastas, and Krause (1974), and Ingle (1973).

3
Economic and Institutional Issues

INTRODUCTION

From a technological perspective, great strides have been made in developing recycling processes that are applicable to many plastic wastes. However, the degree to which these recycling technologies will penetrate the marketplace will largely depend on favorable economic conditions. This chapter abstracts, for the most part, from the purely technical factors that impact recycling and disposal and focuses on the identification and study of the numerous economic and institutional issues that will influence our future decisions to recycle or dispose of plastic wastes.

In most cases the chapter is more conceptual than empirical. This is an unfortunate necessity that stems from the current limited activity in plastics recycling. Some plastics recycling operations are, however, currently active and from them valuable information can be obtained. Important insights can also be drawn from examination of economic and institutional incentives and barriers that exist in other recycling markets. The arguments and methodologies that have been used to evaluate those markets are reviewed and their applicability to current and potential plastics recycling markets assessed.

The discussion is presented in three main sections. The second section of this chapter focuses on the economic and institutional incentives and barriers that face private-sector decision makers when considering plastics recycling. The section begins by identifying the different actors in the private sector that have an impact on whether plastics are recycled or disposed of. Various economic and institutional barriers are then discussed that may affect each set of actors. An important aspect of this section is its discussion of the interdependence of the decisions of the different actors.

The third section of the chapter addresses the public sector's involvement in the recycling question. There are two ways in which the public sector is an active participant in the decision to recycle or dispose of waste: through its selected treatment of plastics in the municipal waste stream, and through legislation and regulations that impact the private sector's decision to recycle. In the case of legislation and regulations that concern the private sector, two important questions arise. First, why should the government take actions to influence the recycling decisions of the private sector? In answer, some have argued that market imperfections exist that cause the level of recycling by the private sector to be lower than the social optimum. Government intervention to promote recycling may therefore be justified on efficiency grounds. It has also been argued that existing government legislation and regulations discriminate against recycling activities. One correction for this alleged discrimination is additional government actions to support recycling. These various arguments are applied to the case of plastics recycling and evaluated from the perspective of economic efficiency. The second question is, given that there are reasons for government to promote additional private-sector recycling, how can that recycling be encouraged? Several government actions are suggested that would stimulate recycling at various levels. These actions are based on the economic and institutional factors that impact the private sector's decision to recycle.

A brief discussion of the economic and institutional factors that impact plastics recycling in other developed countries is presented in the fourth section. The evaluation of recycling activities in these countries is interesting because, in general, their emphasis on plastics recycling is greater than in the United States. This additional interest may be traced to incentives and barriers that differ from those of the United States. Conclusions are summarized in the final section.

It is not the goal of this chapter to make conclusions about the ultimate economic feasibility of plastics recycling. Given our current state of knowledge, those conclusions are not possible. Rather, the goal is to identify the relevant economic and institutional questions to be asked and understand how the incentives and barriers of one set of actors are related to the incentives and barriers of another set of actors and to society as a whole.

THE PRIVATE SECTOR

Identifying the Actors

A useful place to begin our discussion of the incentives and barriers relevant to the private sector is the identification of the actors involved and the interdependencies of the actors' decisions to dispose of or recycle waste products.

Figure 3.1 gives a schematic of how the different segments of the private sector are involved in the decision to dispose of or recycle plastic wastes.

The generation of plastic wastes begins with a) plastics manufacturers, e.g., resin producers, fabricators, converters, assemblers, packagers, and distributors, and b) manufacturers of products containing plastic parts. A manufacturer contributes to the generation of plastics wastes in two ways. First, each manufacturing step entails the generation of a certain quantity of plastic waste, which the manufacturer has the option to either dispose of or

FIGURE 3.1.   The Private Sector's Decision to Recycle or Dispose of Plastic Wastes

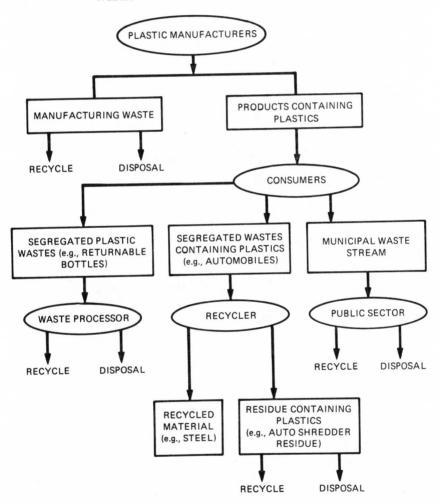

recycle.[1] As was discussed in Chapter 2, manufacturers have commonly recycled a large percentage of their uncontaminated wastes. The remaining
wastes, which have been termed nuisance plastics, are typically disposed by
landfill or incineration, and it is with these nuisance plastics that this study
is predominantly concerned. Second, the plastics manufacturer and the manufacturer of products containing plastics determine the degree to which plastics
can be recycled by designing their products in a way that either facilitates or
hinders their recycling. As is discussed in more detail below, the design of
a product—e.g., the ease of identifying particular resins in products or separating a plastic material from a nonplastic material—will influence the decision to recycle by other actors in the process.

   The consumer of the plastic product is the second actor in the decision
to dispose of or recycle plastic wastes. Once a product that contains plastic
resins becomes waste, the consumer typically disposes the product in the
municipal waste stream or diverts the waste to some alternative stream where
contamination of the plastics with other wastes can be avoided. The consumer
may divert the plastic wastes from the municipal waste stream because the
plastics are contained in a product that is recyclable for its other materials,
such as steel and other metals in junk automobiles. In this case there is an
indirect incentive to divert the plastics from the municipal waste stream based
on the direct incentive to recycle the metals in the junk car. The plastic
components in the recycled product may have no value, given the typical
process by which the product is recycled, and may, in fact, reduce the value
of the product as a recyclable commodity. Alternatively, the consumer may
divert the waste plastic from the municipal waste stream because of a direct
incentive to do so. Returnable deposits on plastic beverage bottles are one
example of a direct incentive to divert the waste from the municipal waste
stream.

   If the consumer does not divert the waste plastic from the municipal
waste stream, the question of recycling or disposal is posed at a public level.
As discussed in Chapter 2, once plastics enter the municipal waste stream,
the difficulty in separating plastics from other waste severely limits the potential for recycling plastics in a relatively uncontaminated form. Tertiary and
quaternary opportunities do, however, exist for the recovery of fuels and basic
chemicals from the assorted municipal refuse. The specific incentives and
barriers faced by the public sector are addressed in the following section of
this chapter.

   If the waste plastic is diverted from the municipal waste stream, the
decision to recycle or dispose of the waste is given to the processor of the
segregated waste or to the recycler of the product containing the waste plastic.
The economic and institutional incentives and barriers faced by these private-
sector actors are similar to those faced by the plastic manufacturer when
making a decision concerning manufacturing wastes.

   In the following subsections the economic and institutional incentives and barriers faced by each of the private-sector actors are reviewed in greater detail.

The Plastics Manufacturer as a Producer of Manufacturing Waste

The condition under which the plastics manufacturer will select recycling rather than disposal can in one sense be stated quite concisely and accurately. There will be an economic incentive for manufacturers to recycle plastic wastes when the marginal cost of recycling realized by the manufacturer is less than the marginal costs of disposal.[2] It is not essential that the revenues from the recycling operation exceed the costs of that operation. Given that the Resource Recovery and Conservation Act of 1976 mandates the abolition of open dumping and requires that all solid wastes be disposed by environmentally sound methods, the plastics manufacturer must either recycle plastic wastes or pay a nontrivial price for environmentally sound disposal methods.[3] A competitive market structure will force the manufacturer to adopt the least costly disposal or recycling method — least costly from the perspective of the manufacturer and not necessarily from the perspective of society.

There are, however, a number of factors that complicate the manufacturer's assessment of the costs and benefits of recycling versus disposal. On the one hand, a decision to recycle may involve a great deal more uncertainty for the manufacturer than that of disposal. Given that manufacturers can be assumed to be risk averse, manufacturers will require a risk premium for selecting a more uncertain over a less uncertain option. Therefore, one cannot simply compare the expected accounting cost of recycling with that of disposal and draw a conclusion about the economic feasibility of recycling. For the manufacturer, recycling will be economically feasible only when the expected cost of recycling plus the risk premium associated with that alternative is less than the expected cost of disposal plus the risk premium associated with the disposal option. Depending on the size of the risk premium required by the potential investor considering recycling, disposal may be selected even though the expected cost of recycling is actually less than that of disposal. This is not, of course, to suggest that disposal does not have its associated uncertainties. The uncertainties of incineration and landfill are, however, more public than private in nature, for example, the potential for air and water pollution. In most cases, these potential external costs are not borne by the manufacturer, but rather are borne by society in general.

On the other hand, factors that may not be obvious to a cursory examination of the problem may make the option of recycling relatively more beneficial than the avoidance of the "out-of-pocket" cost associated with disposal. While these benefits will generally be realized more at the industry level than at the level of the firm, there will be an incentive for individual manufacturers jointly to take advantage of the benefits.

Consider first the potential barriers to recycling posed by the uncertainties of recycling compared with those of disposal. Recycling poses three major types of uncertainty for the manufacturer: technological uncertainties, market uncertainties, and regulatory uncertainties. Technological uncertainties exist because many of the recycling processes have not been proven on a large scale. Information about the technical parameters and operational costs of a particular process are often available only from the developer of the process. Further, the physical and chemical characteristics of the recycled products to be produced from the waste plastics may not be understood well.[4]

Market uncertainties arise from a variety of sources. For example, a great deal of uncertainty exists about the future availability of plastic wastes of particular types and qualities. Firms may therefore be hesitant to invest in large recycling operations, which are required to meet minimum scale requirements, because of fears that feedstocks may not be available. In some cases, especially with many secondary recycling operations, it will be difficult to establish markets for the recycled goods. As discussed in the previous chapter, products from secondary recycling will not usually compete with products produced from virgin resins, but rather will compete with products typically produced from wood and concrete. Potential buyers of recycled goods may be biased against those goods because of uncertainties about their properties or simply because they hold a bias against goods produced from recycled materials, especially recycled plastics. The establishment of distribution channels for waste plastics and recycled products may also be difficult, especially as compared to more traditionally recycled materials such as metals and paper. Recyclers of waste paper and metals produce a feedstock for producers of basic paper and metal products and therefore have a readily identifiable demand source for their intermediate products. In the case of plastics recycling, the potential demand for the recycled products is not so easily identified, thus requiring a greater marketing effort on the part of the recycler. This is especially true in the case of secondary recycling. Tertiary processes that produce intermediate products, such as basic chemicals and fuels, and quaternary processes that retrieve heat energy from plastics may face fewer distribution and marketing problems, giving those forms of recycling a relative advantage over secondary recycling.

The technological and market uncertainties facing manufacturers and firms contemplating recycling manufacturing waste are often compounded by government regulatory uncertainties. Regulations can impact the decision to recycle by making recycling more or less favorable to the private sector as compared with disposal. Regulations that restrict disposal practices in effect increase the cost of disposal and therefore reduce the relative cost of recycling, all else remaining constant. Direct or indirect assistance to recyclers may be provided by other regulations and government actions. Further, regulations and government actions can impact the availabilities of waste, the prices of wastes and recycled products, required pollution control devices, safety requirements, and so on. Unfortunately, government regulatory changes are difficult to predict and therefore contribute to the uncertainty the manufacturers must face. In a paper by Sattin (1978), 41 government laws, regulations, or policies were identified as barriers to the use of secondary metals. Five of the barriers, some of which have applicability to the plastics recycling question, were deemed to be "high priority". Government legislation, regulations, and policies that affect the decision to recycle plastics are discussed in more detail in the following section.

While recycling obviously poses significant uncertainties for the manufacturer, potential benefits are also possible that may not be obvious from a cursory examination of the accounting costs of recycling versus disposal. These benefits are mainly in the form of the goodwill of the consumers of the products produced by the manufacturer, and of the government decision makers who can, through legislation and regulations, seriously affect the economic well-being of the plastics industry. It can be generally concluded that there is a public perception that, aside from the issue of economic efficiency, recycling is a "good thing".[5] There is also a general perception that plastic products are not recyclable. Given these perceptions, plastics manufacturers can establish significant goodwill for their industry and for their products by demonstrating that the technology for plastics recycling exists and in some cases is economically viable. The following subsection considers these issues further.

Table 3.1 summarizes the potential incentives and barriers faced by manufacturers as producers of plastic manufacturing wastes.

Manufacturers of Products Containing Plastic Parts

An equally important set of economic and institutional issues for manufacturers and the degree to which plastic wastes are recyclable concerns the design of products that contain plastic parts. The design of a product may greatly influence the cost of recycling and thus may be the determining factor in an evaluation of the economic viability of recycling. In this subsection we discuss incentives and barriers that affect a manufacturer's decision whether to incorporate recycling objectives in product design. We also examine the incentives for a manufacturer to assist in the design, manufacture, and implementation of plastic-recycling technologies.

The degree to which manufacturers have an interest in the recyclability of their products will be a function of the way in which the products they manufacture are disposed of or recycled. Referring again to Figure 3.1, the consumer of plastic goods typically disposes of those goods in the municipal waste stream or diverts the plastic wastes from the municipal waste stream as a result of a direct incentive to segregate the plastic wastes or an indirect incentive to recycle other components of the products in which the plastics

TABLE 3.1. **Potential Recycling Incentives and Barriers — Manufacturers as Producers of Manufacturing Plastics Wastes**

Incentives

Avoidance of disposal costs
Goodwill of consumers of manufacturer's product
Goodwill of government decision makers who through legislation and regulations can affect the availability and costs of disposal and recycling options and can generally impact the economic viability of the plastics industry.

Barriers

Technological uncertainties of new and often untried recycling processes
Market uncertainties
    Availability of suitable waste in sufficient quantities
    Potential bias against recycled goods by consumers
    Potential difficulty in establishing distribution channels
Regulatory uncertainty
    May impact relative costs of recycling and disposal
    May directly or indirectly exacerbate technological and market uncertainties

occur. The two alternative streams — the municipal waste stream, or a recycling or disposal stream diverted from the municipal waste stream — incur different costs for the producer of that waste and this, in turn, influences demand for the manufacturer's product. On the one hand, if the consumer simply disposes of the consumed plastic product in the municipal waste stream, the consumer faces a zero marginal cost of disposal. In virtually all cases the consumer pays one fixed price for municipal waste disposal, implying that the marginal cost of disposing an additional unit of plastic waste is zero for the consumer. In this case, the cost of disposal will not be an important consideration at the individual level, other than through the individual's concern about the marginal external costs imposed by that disposal. An example of this type of waste is consumer product packaging. On the other hand, if the plastic waste is diverted from the municipal waste stream, the processor of that waste will in most cases face a positive marginal cost of disposal. Plastics contained in scrap automobiles and electrical equipment (which are often recycled for their metal contents) are an example. Beverage bottles that must be returned for a deposit are another example. In these cases, the cost of disposal or recycling will be an important consideration to the consumer of the product because those costs will be directly or indirectly borne by the consumer. It is therefore in the interest of the manufacturers of plastic products that are diverted from the municipal waste stream to reduce the cost of their recycling or disposal. By doing so the manufacturer reduces an indirect cost of the product, i.e., the cost of disposal or recycling, and therefore increases aggregate demand for the products.[6] In addition, the manufacturer by promoting recycling gains goodwill from both consumers and government regulators. The manufacturer must, of course, balance these benefits from promoting recycling against any additional cost of product design and manufacture and the cost of developing recycling processes.

The manufacturer determines the degree to which a plastics product may be recyclable through three important product design decisions: a) the ease with which the plastic components can be identified by resin type; b) the ease with which the plastic components can be dismantled from other components in the product; and c) the types of resins used in the product's manufacture (certain resin mixtures do not recycle well). An example of efforts in this area is the labeling of plastic parts in the manufacture of telephones (see Chapter 6 for details).

The manufacturer can also impact the cost of recycling by developing recycling processes. There are numerous examples of such efforts. The Goodyear Tire and Rubber Company has been active in developing processes to recycle beverage bottles made from polyethylene terephthalate (PET). Goodyear is a major producer of PET [see Goodyear (undated), *Journal of Commerce* (1980), or *Chemical and Engineering News* (1981) for more details]. The Ford Motor Company has been active in the study of recycling plastics from auto shredder residue [see, for example, Mahoney, Braslow, and Harwood (1979) and Harwood (1977a and 1977b)]. More recently, several large plastics producers have joined to form the Plastics Recycling Institute, headquartered at Rutgers University in New Brunswick, New Jersey. The institute, which is a part of the Society of the Plastics Industry, will work on various technical aspects of plastics recycling systems. [See *Chemical and Engineering News* (1985) and *Chemical Marketing Reporter* (1985) for more details.][7]

Table 3.2 summarizes the incentives and barriers for manufacturers to promote the recycling of their products.

Consumers

The consumer plays a key role in determining whether plastic wastes will be recycled or disposed in that it is at the consumer level that many postconsumer plastics either enter the municipal waste stream or are diverted to a recycling stream. As stated earlier, once plastic wastes are contaminated with other wastes in the municipal waste stream, there is little potential for separation and recycling in the secondary sense.

Consumers must dispose, in one way or another, of two types of wastes: a) wastes that normally enter the municipal waste stream and therefore have

TABLE 3.2.   Potential Recycling Incentives and Barriers—Manufacturers as Producers of Products Containing Plastics

Incentives

Reduce the cost of recycling (through product design and development of recycling processes), thereby indirectly reducing the cost of their products to consumers and thus spurring product demand

Goodwill of product consumers and government regulators

Barrier

Potential higher cost of production and the cost of recycling-process development

a zero marginal cost of disposal for the consumer; and b) wastes that cannot
be disposed in the municipal waste stream (such as automobiles and large
appliances) or wastes that through their disposal in the municipal waste stream
impose an indirect cost on the consumer (such as deposit bottles that have
a positive value to the consumer when diverted from the municipal waste
stream but zero value in the municipal waste stream). In the case of the second
type of waste, the consumer must directly or indirectly pay a positive marginal
cost of disposal.

Consider first those wastes that normally enter the municipal waste
stream. Aside from the incentives that result from social consciousness about
the potential external costs imposed by waste disposal, the consumer will have
an incentive to divert those waste materials from the municipal waste stream
only when they have some positive marginal value to the consumer. (Recall
that in the case of manufacturing waste, manufacturers have an incentive to
recycle when the net marginal cost of recycling is less than the marginal cost
of disposal, and not only when the waste plastics have some net positive
value.) While there are reasons for providing a zero marginal cost of disposal
to consumers (see the following section), a costless municipal disposal alternative for consumers may result in some plastic wastes entering the municipal
waste stream when they could have been processed at less cost through a
plastics recycling program. Barriers to the recycling of wastes by consumers
include the inconvenience of source separation and the cost of transporting
the waste to a collection center.

While the barriers to recycling will be similar in the case of the second
type of waste, the consumer will be more inclined to divert the waste from
the municipal waste stream because of the direct or indirect positive marginal
cost of disposal. The consumer's decision to divert the waste will, of course,
be dependent on whether the direct and indirect costs of disposal exceed the
cost and inconvenience of waste diversion. Table 3.3 summarizes the recycling
incentives and barriers faced by the consumer.

Recyclers and Waste Processors

The incentives and barriers faced by a) recyclers of products containing plastic components that must be disposed of as a residue of the recycling operation and b) processors of plastic wastes that have otherwise been diverted from the municipal waste stream are similar to those faced by any product manufacturers deciding whether to dispose of or recycle their manufacturing waste. See Table 3.1 for a summary of those economic and institutional incentives and barriers.[8]

THE PUBLIC SECTOR

The public sector is involved in the decision to recycle or dispose of plastics wastes in two ways: first, through the decisions of local, state, and federal governments to recycle, in one form or another, the plastics that appear in the municipal waste stream; and second, through government legislation and regulations that impact the private sector's incentives and barriers to recycle. In this section we discuss the various economic and institutional factors that must be considered by government in both roles. Throughout the discussion it is assumed that the appropriate criterion for developing and implementing a government role is economic efficiency.

The Public Sector as a Waste Processor

The current typical municipal waste stream contains about six percent plastics by weight. The public sector is faced with the decision of disposing of this waste with other municipal wastes by incineration or landfill, recycling the

TABLE 3.3. Potential Recycling Incentives and Barriers — Consumers

---

Incentives

Direct incentive to divert the plastic waste from the municipal waste stream to a recycling stream and thus claim the positive value of the plastic waste — e.g., returnable beverage bottles

Indirect incentive to divert the plastic waste (which may have no positive value) from the municipal waste stream to recycle other materials in the product containing plastics — e.g., electrical equipment, which is recycled to retrieve its copper content

Concern about potential external cost imposed by the disposal of plastic wastes

Barriers

May be costly and inconvenient to separate plastic wastes

Cost and inconvenience of transportation to collection point

---

plastics wastes along with other wastes in a tertiary or quaternary process, or separating the plastic waste from other municipal wastes and recycling the plastics as a segregated recycling stream. Chapter 2 has discussed the major technological issues to be considered in this decision and has concluded that, in general, the separation of plastics from other municipal wastes, such as paper, is very difficult technologically and will probably not be a viable economic alternative in the foreseeable future. The question faced by the public sector is therefore whether to dispose of waste plastics or to recycle them along with other wastes in a tertiary or quaternary process.

Two key issues must be addressed here. First, what are the direct net costs associated with the two alternatives; and second, what are the external costs associated with the alternatives? The appropriate government decision from an efficiency perspective should, of course, minimize the sum of the two costs. The discussion of the first issue is postponed to Chapter 5 of this book. In general, municipal recycling operations have often "looked good on paper," but have failed to meet their preimplementation economic expectations. A discussion of the external environmental costs of waste disposal and recycling was given in Chapter 2 of this book. Recall that currently there is disagreement about the extent to which plastics and other municipal wastes pose environmental problems when disposed by landfill and incineration. Further, the recycling of plastics and other municipal wastes will not eliminate all environmental damage. The recycling processes may themselves be polluting. Suffice it to say, therefore, that the question of recycling in the municipal waste stream requires additional technical and economic research.

The Public Sector as a Regulator of Private Sector Waste

Given that it is technologically difficult and not economically viable to separate plastics from other municipal wastes, a public decision to promote plastics recycling in a relatively uncontaminated form must be implemented at the level of the private-sector decision maker. In other words, the recycling of plastics in an uncontaminated form must occur outside of the municipal waste stream. In this subsection we discuss why the public sector may, from the perspective of economic efficiency, have a legitimate interest in changing the private sector's waste disposal decisions. Also discussed are various measures the public sector can take to influence the level of recycling by the private sector.

Why Public Involvement?

There are two sets of reasons for potential government involvement in the decisions of the private sector to recycle or dispose of their plastic wastes: a) potential market failures, and b) government regulations that distort the private sector's evaluation of recycling and disposal. In both cases government may be able to influence private-sector decision making and therefore promote a more efficient allocation of resources.

Potential Market Failures. A fundamental result of economic theory is that in the absence of some form of market failure, a competitive market allocation of resources will maximize economic efficiency. Economic efficiency has to do with obtaining the greatest value of real output for a given input cost, or (alternatively) achieving a given value of real output at minimum input cost. If it is given that no market failures exist, government intervention in the market place cannot increase the total economic efficiency of the economy by redistributing resources.[9] Therefore, in order for government intervention in the plastics recycling question (or, for that matter, any intervention) to produce an allocation of total resources that is more efficient than that of the private sector acting alone, some form of market failure must be shown to exist.

In the case of plastics recycling it can be argued that externalities — a form of market failure — result in a private-sector allocation of waste resources and other resources required to recycle waste materials that is suboptimal from the perspective of economic efficiency.[10] These externalities take several forms. The first and most obvious externalities are environmental costs caused by waste disposal, for example, air and water pollution and the disruption of scenic natural environments. It can be argued that recycling reduces these social costs and therefore government promotion of recycling is warranted. However, as discussed earlier in this chapter and in more detail in Chapter 2, the environmental impacts of plastic waste disposal versus those of recycling are not firmly established. Therefore, this market failure remains somewhat dubious.

It has also been argued that materials recycling, in general, prevents some of the environmental damage associated with the production of virgin materials by substituting the production of virgin materials with recycled ones. However, in the case of plastics recycling, this argument is difficult to follow because, unlike other materials such as steel, recycled plastic products will seldom compete with products made from virgin resins. Recycled plastics have numerous potential uses, thus possibly substituting for many different materials — steel, wood, concrete, fuel, and basic chemicals. Further, as discussed earlier, the recycling of plastics will most probably increase, rather than decrease, the production of virgin resins — and their associated environmental costs — by, in effect, reducing the costs of the plastic products to consumers. Therefore, in the case of plastics recycling we must ask which virgin materials are being replaced by the recycled plastics and what are the environmental damages associated with the production of the virgin materials the recycled plastics are replacing? In addition, the analysis must consider that most recycling processes themselves pose various environmental hazards.

A more defensible reason for government intervention in the recycling question concerns the typical subsidization of municipal postconsumer waste disposal. Because households do not normally pay the marginal cost of waste collection and disposal, a divergence will exist between the private and public costs of disposal. In some cases, the total cost of postconsumer waste disposal is paid by general tax revenues. These public policies result in the disposal of larger quantities of solid wastes of all types as compared to the case where marginal disposal fees equal to the social cost of waste disposal are paid by consumers.

An alternative to accepting this market failure is to impose a marginal cost of disposal on consumers. However, implementing such a system would be a costly, if not infeasible, task. Problems would arise in metering and enforcing payment for a marginal contribution to a waste stream at the consumer level. Further, it can be argued that the adoption of such a policy would result in greater, rather than less, environmental damage. A positive marginal cost of waste disposal would make littering a more attractive option for consumers, which, of course, would involve greater environmental damage than disposal through the municipal waste stream.[11] The market distortions caused by the zero marginal disposal cost faced by consumers seem a necessary evil, which, in turn, calls for some form of government recycling assistance. These conclusions do not, however, apply to manufacturing wastes or to wastes that have been diverted from the municipal waste stream. In these cases, the manufacturer or the processor of the diverted waste will typically be forced to pay a marginal cost of disposal (which may or may not include external costs). Further, this argument is also applicable to other wastes that enter the municipal waste stream. The recycling of other materials in the municipal waste stream, such as paper, glass, and metals, faces the same market distortion.

There are three additional but less significant reasons for government intervention on efficiency grounds. First is the "infant industries" argument associated with any new process that has significant social benefits, including, but not exclusively, the prevention of environmental degradation. These benefits may be macroeconomic as well as microeconomic in nature. Second, government has a role to play in the research and development of new recycling processes that have potential payoffs that are high risk and long term. Third, to the extent that plastics recycling reduces the importation of energy, the "oil import premium" suggests that recycling should be subsidized. There is general agreement that a reduction in oil imports provides a social benefit because oil imports are subject to significant supply disruptions and thus large price variations. Of course, there is no reason that plastics recycling should be given preferential treatment over any other measure that might reduce imports.[12]

Regulations That May Hinder Recycling. Several authors have argued that recycling has in general been detrimentally affected by government legislation and regulations. This subsection summarizes these arguments and assesses their validity when applied to plastics recycling.[13]

Discriminatory tax policies are mentioned most often. It is alleged that state tax policies, such as percentage depletion allowances for mining and the production of oil, gas, coal, and sometimes timber, have given the production of virgin materials a significant advantage over recycled materials. Similar tax treatment does not apply to recycled materials.

Some have argued that railroads and other transportation modes have charged higher rates for recycled materials than for virgin materials. While these rate differences are not a direct result of government actions, it has been alleged that government has been ineffective in stopping these potential impediments. Given that transportation costs are often a large portion of the total costs of recycling, these rate differences may place recycled materials at a significant disadvantage to virgin materials. According to Porter (1979), the problem is not one of having legislation to prevent such discrimination. Rather, the problem has been one of enforcing the existing legislation. It has historically been difficult to prove that the differences in rates reflect anything other than actual cost differences. Halgren (1980) argues that in the past there were reasons for cost differences due to the small quantities usually involved with recycled materials and products and the difficulties with transporting often bulky and damaging waste materials. (For example, the jagged edges on metal scrap often damaged box cars.) Halgren goes on to argue, however, that recent containerization techniques have made scrap easier to move. Further, new techniques are available to compress the once bulky materials to smaller volumes. Therefore, according to Halgren, the rate disparities between virgin and scrap materials should be removed.

A third form of alleged regulatory discrimination involves the competitive bidding procedures used by government to procure waste disposal services. In most states laws exist that require the lowest bidder for waste disposal to receive the disposal contract. A potential problem with this procedure is that the lowest bidder will not normally consider the external costs associated with their selected waste disposal method. Those external costs are borne by society rather than the waste disposer, resulting in obvious inefficiencies. Such policies may result in landfill being used for disposal (usually the lowest-cost disposal method), when some form of recycling would be equally attractive when the external costs of disposal are considered.

A further problem with disposal is that existing environmental regulations are often not enforced, thus reducing the cost of disposal. According to Halgren (1980), a survey of reclamation professionals concluded that the enforcement of environmental regulations could "eventually eliminate landfilling by making it too costly" (p. 27). Halgren also states that "nearly two-thirds of all land disposal sites do not meet environmental standards" (p. 27). The nonenforcement of existing disposal regulations obviously gives disposal a relative advantage over recycling.

Zoning laws can also favor disposal over recycling. Regulations concerning the acceptable locations of waste disposal sites often apply to recycling collection and processing centers. While in some cases these regulations are appropriate for recycling centers, in other cases collection centers that have few of the problems associated with large disposal facilities are forced to locate far from consumer suppliers of waste. Given that transportation costs are often crucial to the economic health of a recycling center, these zoning regulations may discriminate against recycling.[14]

Sattin (1978) argues that government sponsored research and development projects are directed much more toward virgin materials than to recycling processes. To the extent that this difference exists, recycling will be at a disadvantage to virgin materials.

Finally, several authors have suggested that labeling requirements concerning the use of recycled materials in products are impediments to the public acceptance of recycled goods. Some of the requirements result from state regulations and some from federal organizations, such as the Federal Trade Commission and the Consumer Products Safety Commission. Opponents of these regulations would prefer the current product notices to be worded differently. They argue that consumers often associate recycled goods with inferior goods. Therefore, the product notices should state whether the qualities of goods made with recycled materials are, in fact, inferior to those of products made with virgin materials.

Table 3.4 summarizes the different potential regulatory and legislative impediments to recycling.

The applicability of these arguments to the plastics recycling question varies from argument to argument. From the perspective of welfare maximization, the first-best response is to remove the various laws and regulations that have caused the market distortions in the first place. However, given that the first-best response is not politically or administratively feasible, a second-best response may involve additional government assistance to promote recycling.

In the case of discriminatory tax laws, the major difficulty in prescribing a counterpolicy is identifying which of the current tax policies discriminate against plastics recycling. As stated earlier, products made from recycled plastics could compete with many different materials. The complexity of identifying the tax laws one is wishing to counter and implementing the countermeasure to impact only those plastic wastes that are negatively affected by the current laws may not be a realistic goal. The same kinds of problems exist with discriminatory freight rates. Putting plastic wastes on an even keel with the virgin materials with which the recycled plastics might potentially compete may not be feasible. However, in the absence of removing the regulations that cause the discrimination, some kind of general assistance to promote plastics recycling may be warranted.

TABLE 3.4. Potential Legal and Regulatory Impediments to Recycling

Tax inequities between virgin and recycled materials
Freight rate differences between virgin and recycled materials
Competitive bidding procedures used by governments to procure waste disposal services, which require acceptance of lowest bid (bids will not include environmental
cost of disposal)
Failures to enforce environmental regulations applicable to waste disposal
Zoning laws that may force recycling centers to locate far from waste sources
Greater emphasis of government R&D activities on virgin materials as compared to recycling processes
Regulations that require products containing recycled materials to be labeled as containing recycled materials

The remaining legislative and regulatory impediments to recycling (summarized in Table 3.4) also suggest that recycling, in general, will be less
attractive than disposal. To the degree that these discrepancies exist, additional government action is warranted in a second-best solution. However,
unlike discriminatory tax policies and freight rates, which, if measurable,
would suggest countermeasures of a particular size, the impacts of these
impediments cannot be traced exclusively to the plastics recycling issue. They
apply equally to all recyclable materials.

Potential Measures to Promote Recycling

It is clear from the above discussions of market failures and impediments to
recycling caused by government regulations and legislation that some government assistance to promote plastics recycling is warranted on the grounds
of maximizing economic efficiency. What is less clear is the degree to which
plastics recycling should be encouraged. While we may acknowledge that there
is a discrepancy between the level of recycling provided by the private sector
and the most efficient level of recycling, it is very difficult to measure that
difference.

Further, it is not clear to what extent the recycling of plastics should be
encouraged as a relatively uncontaminated waste stream, as compared to the
recycling of plastics as a part of the municipal waste stream. A decision here
depends on the environmental damage that plastics may or may not cause
in the municipal waste stream and the social value of the plastic wastes in the
municipal waste stream as compared with their value when diverted from the
municipal waste stream. Answers to these questions await further research.

However, assuming that the public sector does want to encourage the
recycling of plastic wastes outside of the municipal waste stream, what meas-

ures are available to promote recycling? In this subsection several public-sector measures are reviewed that could promote plastics recycling at different levels. Drawing from the discussion of the private sector in the second section of this chapter, recycling can be promoted by simply increasing the incentives for recycling and/or reducing the barriers to recycling.

Non-Consumer Producers of Plastic Wastes. This group includes manufacturers that produce plastic wastes and recyclers and processors of waste products that have been diverted from the municipal waste stream. The incentives and barriers that face this group were summarized in Table 3.1. The incentives to recycling are mainly the avoidance of disposal costs. Government can therefore encourage additional recycling by imposing a tax on the disposal of waste, thus increasing the cost of disposal relative to that of recycling, all else remaining equal. Further, a higher disposal cost encourages the manufacturer to develop and adopt new production processes that produce smaller quantities of wastes. However, a significant problem with this type of policy is that it encourages firms to dispose of their wastes by dumping or other practices that impose a large environmental cost.

The other alternative to promote recycling by these private-sector actors is to increase the incentives to recycle by reducing the uncertainties associated with that alternative. Recall that recycling poses three types of uncertainty for firms: technological, market, and regulatory uncertainties. The public sector can reduce the technological uncertainties by providing assistance in the development and demonstration of recycling processes and by identifying the physical and chemical properties of the wastes and the products that can be made from the wastes.

Market uncertainties can be reduced by providing information to the potential recycler about available recycling processes and the quantities and qualities of different types of plastic wastes that will be produced in future years. Information about the characteristics of recycled products can also be made available to potential consumers of recycled goods to dissipate misconceptions or biases about those products. Government can also reduce market uncertainties by procuring products that are made from recycled plastics or by providing direct subsidies to recyclers.

Regulatory uncertainties can be reduced by simply established consistent and nonfluctuating regulations. This is obviously easier said than done because of the political nature of government decisions. More often than not, a regulation will be imposed for equity rather than efficiency reasons. Government regulators should recognize, however, that fluctuating public policies can have a serious negative effect on recycling, especially when combined with the uncertainties posed by the new technologies and the market place.

Table 3.5 summarizes the potential government measures to promote recycling by non-consumer producers of plastic wastes.

TABLE 3.5. Potential Measures to Promote Recycling by Non-Consumer Producers of Plastic Wastes

Place a tax on disposal with the potential problem of encouraging the dumping of waste materials

Reduce technological uncertainties by providing government support for plastics-recycling R&D

Reduce market uncertainties

  Provide information to potential recyclers about available recycling processes and quantities and qualities of future plastic wastes

  Provide information to potential consumers of recycled products about qualities of those products

  Government procurement of recycled products

  Direct government subsidies to recyclers

Reduce regulatory uncertainties by establishing consistent and nonfluctuating regulations

Consumers. Consumers contribute to plastics recycling by diverting to a recycling stream plastic wastes that would normally go into the municipal waste stream. Government can encourage this behavior in two basic ways: a) by making the option of disposal in the municipal waste stream directly or indirectly more expensive; and b) by making the option of recycling easier and less costly through better collection facilities and changes in product design that allow easier segregation of plastics from other household wastes.

Government actions can directly increase the cost of disposal for the consumer by imposing a positive marginal cost on disposal. However, as discussed earlier in this chapter, this option can be ruled out because it would encourage consumers to dump their wastes outside of the municipal waste stream.

The other option is to increase indirectly the cost of disposal for the consumer, which can be done two ways. First, a tax can be placed on plastics-containing products at the manufacturing, wholesale, or retail levels. From an efficiency perspective, this tax should be equal to the direct marginal cost of disposal in the municipal waste stream plus the marginal external costs imposed by the disposal of that waste. In the case of consumer wastes that are normally disposed outside the municipal waste stream, the tax should be equal only to the external costs imposed by that waste disposal. The processor of this diverted waste will normally pay the direct cost of disposal. This type of tax has the desired effect of forcing the consumer of the plastic product to pay the full cost of that product, inclusive of direct and external disposal costs. The total demand for the plastics product is reduced by the tax, which implies that less of the product will ultimately enter the municipal waste stream.

The major problem with this type of tax is that it does nothing to help divert plastic wastes from the municipal waste stream. Once the tax is paid, the consumer must consider that tax as a "sunk cost" and thus the tax has no impact on the consumer's decision to dispose of the waste in the municipal waste stream or divert the waste to some recycling stream. Therefore, the benefits of the tax are limited to its demand-reduction effects.

The second way the government can impose an indirect cost of disposal on the consumer is through a returnable deposit paid by the consumer at the time of purchase. Again, from an efficiency perspective, this deposit should be equal to the direct plus the external costs of disposal in the municipal waste stream. In this type of program, the consumer has an incentive to divert the waste from the municipal waste stream so the deposit will be refunded. However, if the deposit is not large enough to provide adequate incentive to return the waste, the waste enters the municipal waste stream and the consumer of that product has indirectly paid for the cost of disposal, inclusive of the external or social cost of that disposal. It is important that any deposit on plastic products not exceed the direct plus the external costs of disposal. If the deposit exceeds this limit, the consumer will be encouraged to retrieve the deposit by incurring monetary and nonmonetary costs that are above the costs the plastics would impose if they simply entered the municipal waste stream. When deposits are too large, efficiency is not served, and total social welfare is reduced.

The other basic way government can encourage consumers to divert their plastic wastes from the municipal waste stream is to reduce the monetary and nonmonetary costs of that diversion. One way this can be done is to provide better collection services. This may involve community collection centers or curbside pickup of segregated wastes. Another alternative is to make the segregation of plastics from other wastes easier. Labels that indicate the particular resins in a product would facilitate separation by the household. The government also has a responsibility to provide information to consumers about the various pros and cons of plastics recycling. However, as discussed above, the regulator must be careful not to impose regulations or implement collection systems that impose costs greater than the direct and external costs of disposing the plastic products in the municipal waste stream. Such measures would not be cost effective from the perspective of individual consumers or society as a whole.

Table 3.6 summarizes the potential government measures available to encourage consumers to recycle plastic wastes.

Manufacturers as Producers of Products Containing Plastic Parts. Recall from an earlier section of this chapter that manufacturers have an incentive

TABLE 3.6. Potential Measures to Promote Recycling by Consumers of Plastic Goods

Place tax on plastic goods to reflect the direct and external costs of disposal (does not encourage diversion of waste from the municipal waste stream)
Place returnable deposit on plastic products equal to direct plus external costs of disposal
Provide better collection services for waste materials
Encourage product designs that facilitate segregation of waste plastics
Provide information about the pros and cons of plastics recycling

to design products and processes to facilitate recycling when that design and those processes reduce the cost of disposal for the consumer. By reducing the cost of disposal, the manufacturer indirectly reduces the cost of their product and thus promotes its sale.

Public policies can impact the decisions of the manufacturer to assist in the postconsumer recycling of their products in direct and indirect ways. First, the government can directly require that certain product specifications be met that facilitate recycling. Second, the government can indirectly encourage manufacturer participation by altering the disposal opportunities of consumers and waste processors. The more costly waste disposal becomes for the consumer (due to regulations that encourage the diversion of waste from the municipal waste stream) and for the waste processor (due to, for example, taxes on disposal activities or more restrictive disposal requirements), the more the manufacturer stands to gain from a more easily recycled product.[15]

Examples of Government Legislation to Promote Recycling

There is a multitude of legislation and regulations that impacts the decisions of the private sector to either dispose of or recycle their plastic wastes, and it is not the goal of this section to review or summarize all those legal incentives and barriers. Such a review or summary is beyond the scope of this book. However, it is interesting and worthwhile to examine some recent pieces of legislation, both federal and state, that indicate our growing concern about the environmental costs of disposal and a growing resolve to encourage recycling.

At the federal level the major piece of legislation that impacts private-sector recycling is the Resource Conservation and Recovery Act (RCRA) of 1976, especially subsection D. The RCRA supercedes the Solid Waste Disposal Act of 1965. According to Wakefield (1980), the objectives of the RCRA are to provide technical and financial assistance to waste disposal and recycling, to regulate the disposal of hazardous waste, encourage the production of energy from wastes through resource recovery, and to prohibit the open dumping of wastes.[16] States are required to develop solid waste plans that meet certain minimum requirements stated in the Act. The Act also explicitly requires the Administrator of the Environmental Protection Agency to conduct a scientific, technological, and economic investigation of the potential solutions to the implementation of plastics and glass resource recovery. The RCRA does not, however, require states to recycle plastics or any other material as part of, or outside of, the municipal waste stream. Rather, RCRA "delegates to the states the discretion to develop specific programs to further the Act's broad policy objectives" (Morris, 1981, p. 470). According to Morris, "While most states have enacted the solid waste plans required by RCRA in order to receive federal aid, few states have actively promoted recycling, and fewer still have enacted coherent, systematic recycling programs" (p. 470). Nonetheless, we can conclude that RCRA has been a step toward increasing the incentives for states to recycle all recyclable materials by making disposal a more costly alternative and by providing assistance in the development of recycling operations.

At the state level we find examples of legislation that are much stronger in their encouragement of recycling and have specific implications for plastics. The most noted examples are the beverage-bottle-deposit laws that have been enacted in at least nine states. In most cases the laws require a deposit of five cents on each bottle. These laws have greatly facilitated the recycling of PET, from which most plastic beverage bottles are made. Chapter 6 of this book examines PET recycling in detail.

However, the most ambitious recycling legislation to date is the Oregon Recycling Opportunity Act of 1983. Oregon has historically been in the forefront of recycling legislation, and this legislation continues that bold and potentially costly approach to recycling. The Oregon law requires that every person in the state be given the "opportunity" to recycle. According to Parker (1985) the law requires "(1) a recycling depot with each solid waste disposal site; (2) monthly "curbside" collection of source separated materials within the urban growth boundary of cities of more than 4,000 population, or within the urban growth boundary of a metropolitan service district; and (3) a public education and promotion program" (p. 391). An important part of the legislation is its definition of recyclable material. A recyclable material is "any material or group of materials which can be collected and sold for recycling at a net cost equal to or less than the cost of collection and disposal of the same material" (from the legislation as quoted in Parker, 1985, p. 393). Costs of collection include salaries, transportation cost, equipment cost, and overhead expenses. Note that the definition of recyclable material does not take into consideration the external environmental cost of disposal. Further, note that the legislation allows materials to be grouped so that the materials with greater recycling value can, in effect, subsidize the recycling of materials of less value. The legislation requires that each household, business, and industry be informed of the available recycling programs. Public education programs that emphasize the value of recycling are also required.

The Oregon legislation is in many ways a commendable move in the right direction — particularly in its definition of what constitutes a recyclable material. By not simply mandating that recycling is in the "public good" at any cost, the legislation has laid an appropriate foundation on which an efficient recycling program can be built. The law does, however, face problems. First, no direct incentives other than "civic duty" are provided to the producer of the waste to source separate. Because the proceeds from the more valuable recyclable materials are to be used to subsidize the recycling of less valuable ones, rather than using those proceeds as a direct incentive for consumers to source separate, the participation rate may not be high. Second, it is not clear if the evaluation of a recyclable material will be done on the basis of the average or marginal cost of waste disposal. If average rather than marginal cost is used, obvious problems arise from an efficiency perspective.

According to Parker (1985), the statewide recycling network will not be in place until July 1, 1986. The Oregon experience will offer an interesting case study of the effectiveness of such programs.

INTERNATIONAL COMPARISONS

Halgren (1980) reports that at the time of his writing the United States converted less than 2 percent of its waste into usable products and energy, while Western Europe converted more than 60 percent of its waste. Further, recall from Chapter 2 that many of the technologies that have been developed to recycle plastics have come from Japan and Western Europe. This divergence of interest, or at least action, between the United States and these countries suggests that Europe and Japan may face a different set of incentives and barriers to recycle. This section contains a brief review of some of the differences between the United States and these other developed nations that affect the decision to recycle.

To a great extent, the incentives and barriers to recycle faced by both the public and private sectors in the United States and in Japan and Western Europe are conceptually equivalent. However, there are significant empirical differences. From the public perspective, there is a great interest in recycling all materials because in Japan and most European locations the total cost of disposal (inclusive of all external costs) is higher than in the United States. The environmental cost of disposal by incineration and landfill are made more severe by densely populated areas. For example, while Japan has roughly half the population of the United States, Japan's size is approximately that of Montana, and much of that area is very mountainous. Further, Western

Europe and Japan are generally more dependent on vulnerable imported materials and energy supplies, which makes the potential energy and materials savings from waste recycling relatively more beneficial than in the United States.

Focusing on plastics recycling, as distinguished from the recycling of municipal waste, we find that plastics compose a larger percentage of the total municipal waste stream than in the United States. For example, according to Huls and Archer (1981), 8.4 percent of Tokyo's waste was composed of plastics at that time. This compares to 4–8 percent in Sweden, 5 percent in the United Kingdom, and 5.3 percent in Holland. According to these authors, plastics in the United States accounted for 3.7 percent of the total waste stream. These factors have contributed to a keen public interest in recycling in Japan and Western Europe, which has resulted in several government policies to encourage resource recovery.

These government policies, in combination with a higher direct cost of disposal, have pushed the private sector to consider recycling. Baller (1982) reports that Japan has launched a national recovery plan, which involves the development and demonstration of technology to recycle municipal wastes. The Japanese have taken the position that pyrolysis will be the most efficient means of recycling plastics and other high-caloric wastes. According to Basta et al. (1984), West Germany has given serious consideration to passage of a law that would require households to keep plastic wastes in separate bins, which would be collected by garbage haulers. In many Western European countries it is common to have bottle laws that require deposits or prohibit the use of nonreturnable containers.

In general, the economic and institutional incentives to recycle are stronger and the barriers weaker in Japan and most Western European countries. This has led to more public and private sector activity in the recycling of plastics and other materials. However, many of the attempted recycling programs have failed for one reason or another. Further, several authors, for example Baller (1982) and Milgrom (1982), believe that many operational foreign recycling programs are not economically viable without significant government subsidization. Whether this subsidization is consistent with the maximization of total social welfare is open to question.[17]

CONCLUSIONS

The discussion in this chapter of the economic and institutional issues that concern the decision to recycle allows us to draw several general conclusions. First, the economic viability of plastics recycling cannot be assessed by simply evaluating the expected net accounting revenues or costs to the recycler. There are numerous economic and institutional incentives and barriers that will impact the private sector's decision to dispose of or recycle their plastic wastes. Moreover, the specific incentives and barriers will differ depending upon the particular private-sector actor — i.e., manufacturer, consumer, recycler, or waste processor.

Second, there are defensible arguments that suggest that the level of recycling provided by the private sector will not be adequate from the perspective of efficiency maximization. These arguments are based on the externalities associated with waste disposal and recycling and the government regulations and legislation that serve as impediments to recycling activities. To the extent that the regulatory and legal impediments cannot be removed because of political constraints and to the extent that externalities distort the optimal social response to the problem, government action to encourage recycling is appropriate.

Third, if recycling is to be encouraged by government actions, those actions should be selected carefully and in accordance with reducing the private sector's barriers and/or increasing the private sector's incentives to recycle. Care must be taken to encourage recycling, rather than simply make the option of disposal more costly, which will promote open dumping. Further, if from a public perspective we desire to recycle plastics outside of the municipal waste stream as a relatively uncontaminated waste, recycling programs should encourage the diversion of plastics from the municipal waste stream at the consumer level. It is generally felt that once the consumer disposes plastics in the municipal waste stream, those plastics cannot be economically separated from other similar waste materials.

Finally, government regulators must be careful when implementing recycling programs not to impose costs on society that exceed the costs society would bear if the waste were simply disposed of by conventional means. Recycling is not a "good thing" in and of itself. Its worth must be evaluated from the perspective of economic efficiency, inclusive of all direct and external costs.

NOTES

1. See Chapter 4 for details on the estimated production of plastics waste from each manufacturing step.

2. It is important that we distinguish between the costs of disposal and recycling realized by the manufacturer or other private sector actors and the total costs of disposal and recycling, inclusive of external costs, realized by society as a whole. While individuals making decisions about recycling or disposal will generally not consider external costs, those costs will, of course, have a great impact on the need for government or public-sector actions to promote recycling.

3. See Chapter 5 for a discussion of the costs of waste disposal by landfill and incineration.

4. See Chapter 2 for a discussion of the characteristics of several recycling technologies now available and in the development process.

5. For a discussion of the public's perception of recycling see, for example, Quade (1982) and Sharpe (1977).

6. There is an interesting difference between plastics recycling and the recycling of more conventionally recycled materials, such as steel, aluminum, and copper. In the case of these other materials, an increase in recycling directly reduces the demand for virgin materials in that recycled materials replace virgin materials. Manufacturers of virgin materials therefore have little incentive to promote recycling. However, in the case of postconsumer plastics, increased recycling will not necessarily reduce the demand for plastic resins, but rather will in most cases increase it by making plastics generally more acceptable. Products that result from secondary, tertiary, or quaternary plastics recycling do not in most cases compete with products manufactured from virgin resins. No displacement of virgin resins usually occurs. Manufacturers of resins and products containing plastics will therefore have a direct incentive to promote plastics recycling.

7. Chapter 6 contains detailed discussions of the recycling of PET beverage bottles and the potential recycling of plastics from automobile shredder residue.

8. Additional discussion of the economic and institutional barriers that may face recyclers of plastic wastes is contained in Bollard (1982), Henstock (1980), Bever (1977), and Arthur D. Little, Incorporated (1973).

9. For a discussion of what constitutes a market failure and the implications of those failures for economic efficiency, see, for example, Layard and Walters (1978) p. 22-26. Economic efficiency must, of course, be distinguished from distributive equity, which has to do with how limited resources are divided among individuals and groups. In the case of plastics recycling, the issue of distributive equity arises if we as a society wish to trade off a degree of economic efficiency so that recycling can be subsidized. Here I shall abstract from equity issues for two reasons. First, I have little to say about equity/efficiency tradeoffs beyond my own individual preferences. Second, analysis of efficiency is essential in measuring the trade-offs that the pursuit of equity goals always entails — in other words, in telling not only whose "pocket is being picked" but also how badly.

10. Other potential market failures — i.e., increasing returns to scale, the lack of a full set of contingent commodities markets, and a noncompetitive market structure — do not have great significance for the plastics recycling question. There is no evidence of significant increasing returns to scale for nonmunicipal waste recycling programs nor any evidence that problems with contingent commodities markets are any more severe than in any infant industry. Further, Baumol (1977) argues that a monopolistic element in either the industry that provides virgin materials or the industries that use those materials as inputs will not contribute to an argument for government involvement, but rather will detract from the virtues of such a program. These industries will likely restrict the production and use of the material to be recycled, thus reducing the marginal benefits of processes that are designed to recycle the materials.

11. See Anderson (1977) and Baumol (1977) for good overviews of the general arguments for and against government intervention to promote materials recycling.

12. See Broadman (1982) for a discussion of the "oil import premium" from both conceptual and empirical perspectives. Broadman reports that estimates of the premium ranged in 1981 from a low of about $8 to a high of about $124 per barrel.

13. This subsection draws from the discussions of legal impediments to recycling from the following sources: Halgren (1980), Sattin (1978), Baumol (1977), Anderson (1977), and Bever (1977). While the arguments presented in these works were not directed to plastics recycling in particular, many of the conclusions reached are relevant to the plastics recycling issue. It is suggested that the interested reader refer to the Halgren and Sattin papers for the most in-depth discussions of the potential problems posed by these legal impediments.

14. See Halgren (1980, pp. 20-22) for more on discriminatory zoning laws.

15. For additional discussion of the general measures available to promote recycling, see,

for example, Morris (1981), Bradley and Florio (1980), Anderson (1977), Baumol (1977), Pearce and Grace (1976) and Arthur D. Little, Incorporated (1973).

16. See, for example, Morris (1981) for further discussion of the Resource Conservation and Recovery Act.

17. The interested reader can obtain additional information about the technical, economic, and institutional aspects of recycling in Japan and Western Europe from the following articles and reports: Basta et al. (1984), Baller (1982), Milgrom (1982), Huls and Archer (1981), Monsanto Research Corporation (1981), Trauernicht (1981), Zalob (1979), and Bidwell and Raymond (1978).

4

Plastic Waste Projections

INTRODUCTION

One of the major uncertainties facing the public and private sectors in their decisions about plastics recycling and disposal concerns the production of future plastic wastes. Recall from Chapter 2 that the applicability of the different recycling technologies depends in part on the resin being recycled and the level to which the waste is contaminated with other waste materials. Further, recall that different resins may pose different environmental implications, especially when incinerated. It is therefore important that information be available about the types and quantities of resins that are expected to enter the waste stream in future years, as well as the form in which those resins will either be disposed or recycled.

The purposes of this chapter are thus threefold. First, the chapter contains a discussion of the historical production and use of plastic resins in the United States by major product category. Projections of total resin production and use by major product category are given through the year 1995. Second, the chapter includes estimates and projections of the quantities of manufacturing nuisance plastics for the years 1984, 1990, and 1995. The estimates and projections are disaggregated by resin type and by major manufacturing process. Third, the chapter contains estimates and projections of the quantities of postconsumer plastic wastes for the years 1984, 1990, and 1995. Estimates and projections are given by major product category and by resin type.

Waste projections disaggregated by manufacturing nuisance plastics and by major postconsumer product categories are valuable in that they suggest the degree to which future waste plastics will be contaminated with other materials. These projections also help to identify plastic waste streams that will be difficult and relatively easy to divert from the municipal waste stream. The applicability of the major recycling technologies—i.e., secondary, tertiary, and quaternary—to different segments of the plastics waste stream can therefore be assessed from a technological perspective.

In each of the following sections, which detail these estimates and projections, a brief description of the methodology used is given. The interested reader is referred to Appendix B for more detailed descriptions of the methodologies. Appendix B also contains more detailed numerical results that for the sake of clarity have been presented in graphic form in this chapter.

HISTORICAL AND PROJECTED
U.S. RESIN PRODUCTION AND USE

The Society of the Plastics Industry (1984) reports that total U.S. production of plastic resins increased from about 8.2 billion pounds in 1962 to about 42.8 billion pounds in 1983—a more than fivefold increase. The rapid growth in the production and use of plastics has been accompanied by the development of numerous new resins and innovative new products that utilize resins by themselves and in composite structures. This section details this rapid growth and presents projections of the U.S. production and use of plastics through the year 1995.

Methodology

Projections of total U.S. resin production and the use of resins in major product categories were made using time series analysis. Historical production of all resins and the use of resins in eight major product categories were regressed against time and a constant. That relationship was then used to project the future production and use of plastic resins in the United States.[1]

Results

Figure 4.1 illustrates the historical growth of total resin production in the U.S. for the years 1962 to 1983 and gives projections of future production through the year 1995. Production increased steadily between the years 1962 and 1973 and again between 1975 and 1979. The rather sharp reductions in production that occurred during the mid 1970s and then again in 1979 and the early 1980s can be explained, at least in part, by the world oil supply disturbances that occurred during those time periods. While there have been downturns in production, the general trend has been sharply upward during the past two decades. Further, given the methodology discussed above, projections indicate

FIGURE 4.1. Total U.S. Resin Production (billions of pounds)

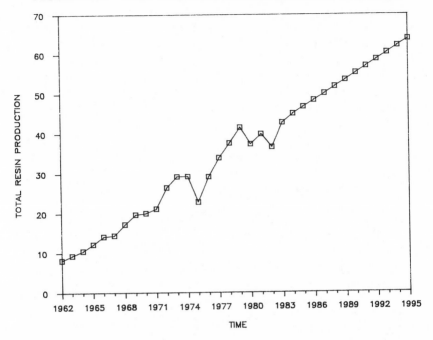

Notes: Historical data 1962–1983 includes the production of polyurethane. Numerical values can be found in Table B.5 (Appendix B).

Source of historical data: *Facts and Figures of the U.S. Plastics Industry*, Society of the Plastics Industry (1984) (used with permission).

that by 1990 total production will exceed 55 billion pounds and by 1995 will approach 64 billion pounds per year.

Figures 4.2a and 4.2b illustrate the historical use of plastic resins in several major product categories for the years 1974 through 1984 and give projections of future plastic use through the year 1995. The Society of the Plastics Industry's *Facts and Figures of the U.S. Plastics Industry* (from which historical data were obtained) excludes the use of polyurethanes in their data series on the use of resins in major product categories. The results reported in Figures 4.2a and 4.2b do not therefore include the use of polyurethanes. (The use of polyurethanes in major product categories is considered below.) Product categories include packaging, consumer and institutional goods, transportation, electrical and electronic goods, building and construction materials, furniture and fixtures, and industrial machinery. (The use of plastics in the "other" category is discussed in Appendix B.)

Given our product classifications, the two largest users of plastic resins are packaging and building and construction materials. The use of resins for packaging has increased from 6.7 billion pounds in 1974 to 12.4 billion pounds in 1984. Projections indicate that plastics in packaging will increase to 16.1 billion pounds by 1990 and to 19.1 billion pounds by 1995. The use of plastics in building and construction materials has also increased sharply in recent years. Plastics in that sector increased from 4.3 billion pounds in 1974 to 9.7 billion pounds in 1984. Projections suggest the use of plastics in building and construction will increase to 12.0 billion pounds in 1990 and further increase to 14.5 billion pounds in 1995. The third largest use of plastics is for consumer and institutional goods. However, this product category has not experienced the large growth in plastics consumption that has been observed in the two largest categories, and it is projected that future increases in the use of plastics in this category will be moderate. The use of plastics in consumer and institutional goods increased from 3.2 to 4.0 billion pounds from 1974 to 1984.

FIGURE 4.2a.   Resin Usage by Product Type (billions of pounds)

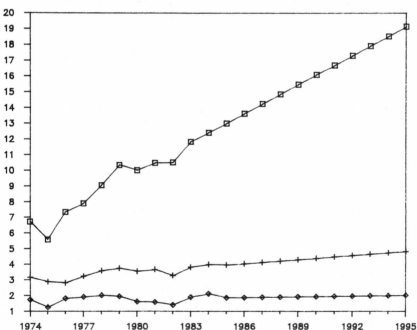

□ Packaging; + Consumer and Institutional Goods; ◇ Transportation

Notes: Historical data 1974–1984. All data exclude polyurethanes. Numerical values can be found in Table B.5 (Appendix B).

Source of historical data: *Facts and Figures of the U.S. Plastics Industry*, Society of the Plastics Industry (1984 to 1985) (used with permission).

FIGURE 4.2b.   Resin Usage by Product Type (billions of pounds)

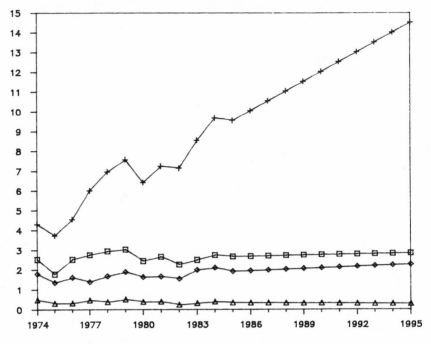

+ Construction; □ Electrical and Electronic Goods; ◇ Furniture and Fixtures; △ Industrial
Machinery

Notes: Historical data 1974–1984. All data exclude polyurethanes. Numerical values can
be found in Table B.5 (Appendix B).

Source of historical data: *Facts and Figures of the U.S. Plastics Industry*, Society of the
Plastics Industry (1984 and 1985) (used with permission).

It is projected that the use of plastics in this category will increase to 4.4 and
4.8 billion pounds in 1990 and 1995, respectively.

The use of plastics in the remaining four major product categories has
been relatively flat during the 1974–84 time period and projections indicate
only small changes in the use of plastics in these categories through 1995. The
use of plastics in electrical and electronic equipment is expected to increase
from its 1984 level of 2.8 billion pounds to 2.9 billion pounds in 1995. The
use of plastics in furniture and fixtures is expected to increase from its 1984
level of 2.1 billion pounds to 2.3 billion pounds in 1995. The use of plastics
in the domestic transportation sector is actually expected to decrease slightly
from its 1984 level of 2.1 billion pounds to 2.0 billion pounds in 1995.[2] Finally,
the use of plastics in industrial equipment is expected to decrease from its
relatively small 1984 level of 0.4 billion pounds to 0.3 billion pounds in 1995.

The historical and projected consumption of polyurethanes in selected markets is graphically displayed in Figure 4.3. By far the largest use of polyurethanes is in the manufacture of furniture. Polyurethane usage in furniture manufacture has increased from 0.7 billion pounds in 1973 to 0.9 billion pounds in 1983. Projections show that usage increasing to 1.3 and 1.5 billion pounds in 1990 and 1995, respectively. The second largest user of polyurethane is in building and construction materials, which has recently overtaken the transportation category. Polyurethane usage in construction is expected to increase from its 1983 level of 0.4 billion pounds to 0.7 billion pounds in 1995. Projections show the use of polyurethane dropping in the domestic transportation sector from its 1983 level of 0.3 billion pounds to 0.2 billion pounds in 1995. Polyurethane use in packaging has recently overtaken the electrical and electronic category and is expected to increase to about

FIGURE 4.3. Polyurethane Foams in Selected Markets (billions of pounds)

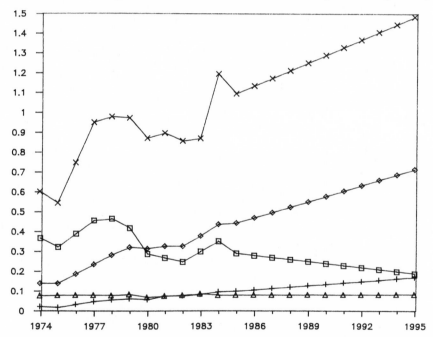

× Furniture and Fixtures; □ Transportation; ◇ Construction; △ Electrical and Electronic Goods; + Packaging

Notes: Historical data 1974–1983. Data entries for 1974 and 1976 have been interpolated by the author. Numerical values can be found in Table B.7 (Appendix B).

Source of historical data: *Facts and Figures of the U.S. Plastics Industry*, Society of the Plastics Industry (various issues) (used with permission).

0.2 billion pounds by 1995 — or about the level of the transportation sector. The use of polyurethane in electrical and electronic equipment is expected to decrease slightly during the next decade.

PROJECTIONS OF MANUFACTURING NUISANCE PLASTICS

Recall from Chapter 2 that manufacturing nuisance plastics are those waste resins produced during various stages of the manufacturing process that are normally disposed of rather than recycled. While much manufacturing waste can be recycled, a significant quantity of waste has historically been disposed of because of contamination or because the resins were difficult to recycle due to their physical and chemical properties. This section attempts to estimate the quantities of different resins produced in various manufacturing stages in the year 1984. Projections are attempted for the years 1990 and 1995. Unfortunately, according to Leidner (1981), our knowledge is somewhat lacking with respect to the generation of waste in different stages of production. The estimates and projections contained in this section — which are based on information in Leidner's 1981 book — must therefore be taken as rough estimates and projections. Leidner (1981) states that progress is being made toward the development of scrapless plastics processing equipment. To the extent that these developments have been implemented, or will be implemented, the estimates and projections given in this section will be biased upward.

Methodology

The methodology to estimate and forecast the quantities of manufacturing nuisance plastics included the following steps. First, estimates of the percentage of total plastic throughout that has historically become a nuisance plastic at various stages of production — i.e., resin producer; fabricator; converter; and packager, assembler, and distributor (hereafter termed distributor) — were obtained from Leidner (1981). Because more recent information was not identified, the estimates from Leidner were assumed to remain constant over the projected time period. Second, the percentages derived from the first step were applied to the projections of total future U.S. resin production discussed in the previous section. This step gives projections of the total quantities of nuisance plastics to be produced by the different manufacturing processes or stages of production. Third, the estimates and projections derived from the second step were disaggregated by resin type according to the percentages of each resin produced in the U.S. in 1983. Estimates and projections are thus provided by manufacturing process and by resin type.

Table 4.1 gives estimates of the percentages of throughput that become nuisance plastic at different stages of production. For a constant level of

TABLE 4.1.   Percentage of Throughput That Becomes a Nuisance Plastic at Various Stages of Production

	Percentage of Throughput That Becomes a Nuisance Plastic	Percentage of Commodity Resin Affected
Resin Producer	1.5	100.0
Fabricator	2.4	84.2
Converter	4.8	40.1
Packager, Assembler and Distributor	1.2	68.0

Source: Leidner (1981), pp. 67–73.

throughput, converters produce the most nuisance plastics, followed by fabricators, resin producers, and distributors. However, not all resins produced are subject to all stages of production. Figure 4.4 illustrates the percentage of nuisance plastics produced by each production stage when adjusted for the percentage of commodity resin affected by each stage. Given this adjustment, fabricators produce 32.3 percent of all nuisance plastics, followed by converters at 30.7 percent, resin producers at 24.0 percent, and distributors at 13.0 percent.

Table 4.2 gives estimates and projections of manufacturing nuisance plastics and disaggregates those estimates by stage of production. The numbers include the production of polyurethane. Manufacturing nuisance plastics are estimated to have totaled about 2.8 billion pounds in 1984. Total nuisance plastics are projected to increase to about 3.5 billion pounds in 1990 and to about 4.0 billion pounds in 1995.

Figure 4.5 illustrates the estimated percentage composition of manufacturing nuisance plastics by plastic type. Thermoplastics are projected to compose 82.2 percent of the total. Thermosets and polyurethane foams are projected to compose 13.3 and 4.5 percent, respectively. Table B.8 in Appendix B contains more disaggregated projections by specific resin. Phenolics are expected to be the most produced thermoset and polyethylene the most produced thermoplastic. Recall that it has been assumed that the composition of waste by plastic type does not change over the time period of the projections.

POSTCONSUMER PLASTIC WASTES

Recall from Chapter 2 that one of the main technological constraints preventing the primary, secondary, and in some cases tertiary recycling of plastic wastes is the difficulty in separating plastics from other waste materials and

FIGURE 4.4. Manufacturing Nuisance Waste Estimates (percentage by manufacturing process)

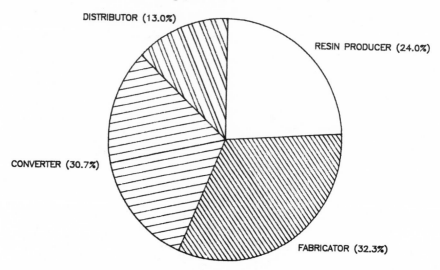

Source: Derived from information given in Leidner (1981), pp. 67–73.

TABLE 4.2. Estimates and Projections of Manufacturing Nuisance Plastics Disaggregated by Stage of Production (in millions of pounds)

	Year		
	1984	*1990*	*1995*
Resin Producer	676	829	957
Fabricator	910	1,117	1,289
Converter	867	1,064	1,228
Packager, Assembler and Distributor	368	451	521
Totals	2,821	3,461	3,995

Source: Compiled by the author.

FIGURE 4.5. Manufacturing Nuisance Waste Estimates (percentage by plastic type)

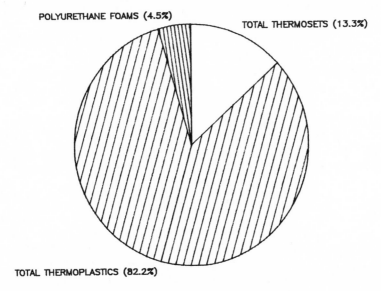

POLYURETHANE FOAMS (4.5%)

TOTAL THERMOSETS (13.3%)

TOTAL THERMOPLASTICS (82.2%)

in separating individual resins. It is generally concluded that once a waste plastic enters the municipal waste stream, recycling opportunities are largely limited to tertiary and quaternary recycling methods. It is therefore important that we have information about the projected levels of postconsumer waste not only in terms of quantity and resin type, but also in terms of the sources of that waste. By identifying the waste sources, we can project the quantities and types of waste that can be diverted from the municipal waste stream and thus become candidates from secondary recycling. We can also project the quantities of plastic waste that, because of the products and waste streams in which they appear, will be very difficult to collect outside of the municipal waste stream. The recycling possibilities for these plastics will probably be limited to tertiary and quaternary recycling as part of the municipal waste stream.

Methodology

Postconsumer plastic wastes have been estimated and projected for eight product categories by resin type for the years 1984, 1990, and 1995. The product categories include transportation, packaging, building and construction, electrical and electronics goods, furniture and fixtures, consumer and institutional goods, industrial machinery, and adhesives and other goods. The methodologies for the different product categories differed slightly because of data limitations. The specifics of the methodologies for each product category are given in Appendix B. However, the general methodology was straightforward and included the following steps. First, information was obtained on the average life spans of the products in the eight product categories. Second, information was obtained on the historical use of specific plastic resins in those product categories. Estimates and projections of the plastic wastes from these product categories can therefore be made.[3]

Information about the average life spans of products in the major product categories is available from several sources.[4] For the vast majority of product categories the estimates do not differ significantly from source to source. The exception is the building and construction category where estimates differ by as much as 100 percent. The most recent information identified on product life spans comes from Leidner (1981) and is used in part to arrive at the numbers presented in Table 4.3. Building and construction materials have the longest life span of all categories considered, at about 25 years. Packaging is at the other extreme, with a life span of less than one year.

Ideally, in addition to knowing the expected life spans of the products, it would also be beneficial to have information about the distributions of the life spans. While Holcomb and Koshy (1984) give this information for the case of automobiles, information of this type was not found for other transportation equipment or for any of the other major product categories. Obvious problems arise if, for example, we are forecasting 1990 plastic wastes

TABLE 4.3. Estimated Life Spans of Selected Products (in years)

Product Category	Estimated Life
Transportation[a]	11
Packaging	< 1
Building and Construction	25
Electrical and Electronics	15
Furniture and Fixtures	10
Consumer and Institutional	5
Industrial Machinery	15
Adhesives and Other[b]	4

Source: Leidner (1981, p. 81) unless otherwise noted
[a]Holcomb and Koshy (1984, pp. 2–18) report that automobiles have median lifetimes of 10.9 years, light trucks 14.9 years, and heavy-duty trucks 12.0 years.
[b]Assumed by the author.

from automobiles (with average lives of 11 years) and 1979 was a particularly depressed year for the domestic automobile industry. To help minimize this type of problem, a three-year average of the plastics consumed in products with relatively long lifetimes was used where the data would permit. (See Appendix B for details.)

Data on the historical consumption of plastic resins by plastic type and by major product category were obtained from various issues of the SPI's *Facts and Figures of the U.S. Plastics Industry*.

Results

Estimates for 1984

It is estimated that about 29.6 billion pounds of postconsumer plastic waste were produced in 1984. Figure 4.6 illustrates the disaggregation of that total into the eight product categories discussed above. Figure 4.7 illustrates the percentage composition of the 1984 total in terms of product type. Packaging is by far the largest product category at about 12.5 billion pounds or about 42.2 percent of the total. Plastic wastes from packaging are estimated to have been composed almost entirely of thermoplastic resins — 98.7 percent of the total. Thermosets contributed 0.5 percent and polyurethane foams 0.8 percent.[5]

The second largest product category is the "other" category at 20.0 percent of the total or about 5.9 billion pounds.[6] The vast majority of this category is estimated to have been composed of thermoplastics at 4.8 billion pounds or 80.4 percent of the total. Thermosets represented 18.0 percent and polyurethane foams 1.6 percent.

Consumer and institutional goods were in third place, representing 12.3 percent of the total postconsumer wastes produced in 1984, and were composed predominantly of thermoplastics at 3.4 billion pounds or 95.2 percent of the total in that product category. Electrical and electronic goods, furniture and fixtures, and transportation equipment represented 8.0, 7.4, and 6.6 percent of total 1984 postconsumer wastes. Wastes from the electrical and electronics category were mainly thermoplastics at about 82.6 percent. Wastes from the furniture and fixtures category were composed of 62.6 percent thermoplastics, 9.9 percent thermosets, and 27.5 percent polyurethane foam. Transportation wastes were composed of 51.7 percent thermoplastics, 29.9 percent thermosets, and 18.2 percent polyurethane foam.

The quantities of postconsumer plastic wastes from the building and construction and the industrial machinery categories were relatively small in 1984 at 2.4 and 1.2 percent of the total, respectively. Wastes from the building and construction category were mostly thermoplastics at 79.6 percent of the total. All but a minute part of the remainder was composed of thermosets.

FIGURE 4.6.   1984 Postconsumer Waste Estimates (billions of pounds)

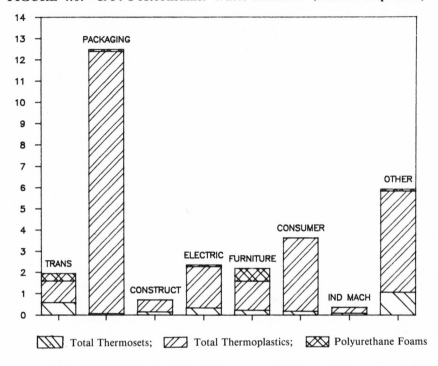

FIGURE 4.7.   1984 Postconsumer Waste Estimates (percentage by product category)

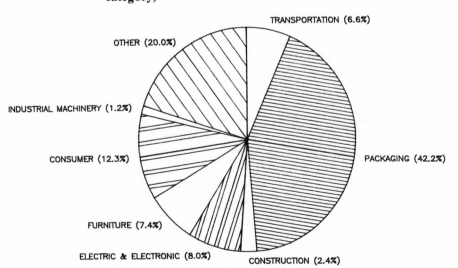

Wastes from industrial machinery were composed of 77.6 percent thermoplastics and 22.4 percent thermosets.

Figure 4.8 illustrates that aggregate postconsumer wastes in 1984 are estimated to have been composed of about 86.9 percent thermoplastics, 9.0 percent thermosets, and 4.1 percent polyurethane foams.

Projections for 1990

Postconsumer plastic wastes are projected to increase from their estimated 1984 level of 29.6 billion pounds to about 35.0 billion pounds in 1990. Figure 4.9 illustrates the composition of the 1990 projection by major product category. Figure 4.10 indicates the percentage that each product category is projected to compose. Packaging is projected to remain as by far the largest single contributor to the postconsumer waste stream at 46.3 percent of the total—up from its 1984 level of 42.2 percent. Virtually all packaging material is projected to be thermoplastics.

Only very slight changes are expected in the remaining product categories. Plastic wastes from the furniture sector are expected to slightly overtake waste production from the electrical and electronics category. All other product categories are expected to contribute slightly less in percentage terms. How-

FIGURE 4.8.   1984 Postconsumer Waste Estimates (percentage by plastic type)

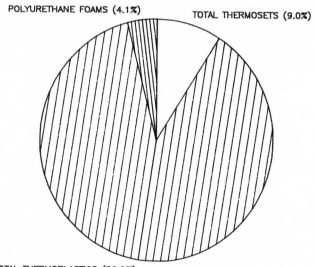

POLYURETHANE FOAMS (4.1%)

TOTAL THERMOSETS (9.0%)

TOTAL THERMOPLASTICS (86.9%)

FIGURE 4.9.   1990 Postconsumer Waste Projections (billions of pounds)

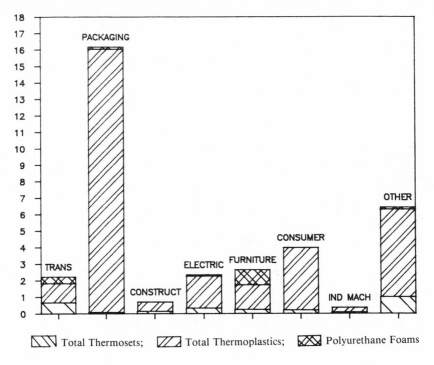

FIGURE 4.10.   1990 Postconsumer Waste Projections (percentage by product category)

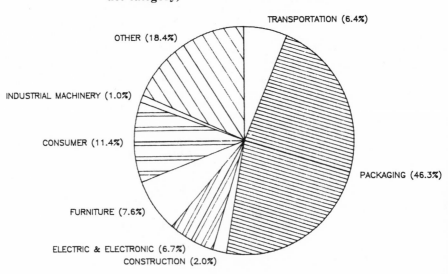

ever, in absolute terms all product categories are projected to produce higher levels of plastic wastes in 1990 than in 1984.

Figure 4.11 illustrates the percentage composition of total 1990 postconsumer waste by plastic type. Thermoplastics are expected to slightly increase their share of the total from 86.9 percent in 1984 to 87.3 percent in 1990. Thermosets are projected to compose 7.9 percent of the 1990 total, as compared to 9.0 percent in 1984. Polyurethane foams increase slightly from 4.1 to 4.8 percent.

Projections for 1995

Total postconsumer plastic wastes in 1995 are projected to be in excess of 43.4 billion pounds — a 24.2 percent increase over the 1990 level and a 46.6 percent increase over the 1984 level. Figure 4.12 illustrates the composition of the 1995 level by product category in billions of pounds. Figure 4.13 presents the composition of 1995 postconsumer plastics by product category in percentage terms. Packaging is projected to remain the largest source of plastic waste at 19.3 billion pounds or 44.5 percent of the total. "Other" goods are projected to compose 16.6 percent of the total, while consumer and institutional goods are projected to be in third place at 10.1 percent of the total.

The most dramatic change projected to occur in 1995 is the flow of postconsumer plastics from the building and construction sector. Recall that

FIGURE 4.11.   1990 Postconsumer Waste Projections (percentage by plastic type)

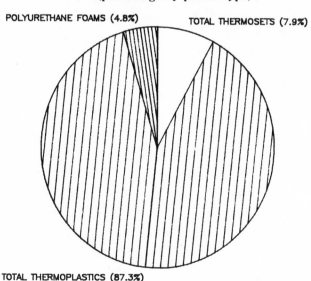

POLYURETHANE FOAMS (4.8%)   TOTAL THERMOSETS (7.9%)

TOTAL THERMOPLASTICS (87.3%)

FIGURE 4.12.   1995 Postconsumer Waste Projections (billions of pounds)

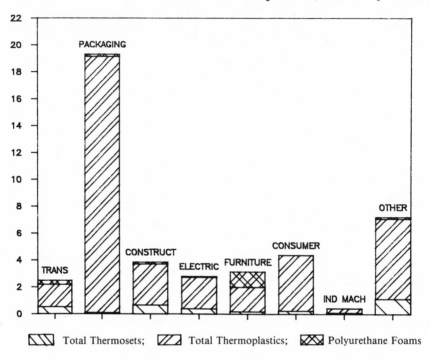

⟨Χ⟩ Total Thermosets;   ⟨Ζ⟩ Total Thermoplastics;   ⟨Χ⟩ Polyurethane Foams

FIGURE 4.13.   1995 Postconsumer Waste Projections (percentage by product category)

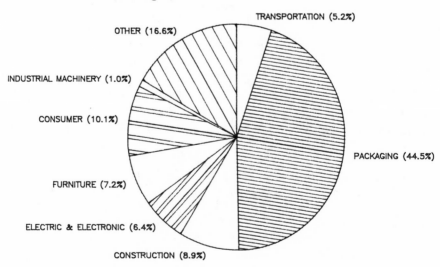

building and construction products are assumed to have average life spans of 25 years. Therefore, the revolution in the use of plastics in building and construction materials in the 1960s and early 1970s will begin to have a significant impact on postconsumer wastes only in the 1990s. Postconsumer wastes in 1995 from the building and construction sector are projected to be 3.9 billion pounds or 8.9 percent of all postconsumer plastic wastes. Recall that this category accounted for only 2.0 percent of the total projected for 1990. Moreover, in years beyond 1995 the total quantity of waste from the building and construction sector is expected to increase sharply as the plastics used in the 1970s and 1980s enter the waste stream. In addition, Figure 4.2 indicates that the growth in the use of plastics in the building and construction sector is projected to be large through 1995 — surpassed only slightly by that in the packaging category. Recall that plastics consumption in building and construction materials is currently exceeded only by packaging.

The total quantities of postconsumer plastics from the remaining product categories are projected to increase slightly in 1995 from their 1990 levels. However, in percentage terms these remaining categories are projected to compose a smaller percentage of all postconsumer wastes. Wastes from the transportation sector are projected to increase only slightly and are projected to compose 5.2 percent of total postconsumer waste as compared to 6.4 percent in 1990.

Figure 4.14 illustrates the projected composition of 1995 postconsumer plastic wastes by plastic type. Thermoplastics are projected to account for 87.8 percent of the total. Thermosets and polyurethane foam are projected to compose 7.6 and 4.6 percent of the total, respectively. This composition varies only slightly from that projected for 1990.

CONCLUSIONS

Figure 4.15 illustrates the projected growth of manufacturing nuisance plastics and postconsumer plastic wastes. The total for the two groups of plastic wastes is projected to increase from its estimated 1984 level of 32.5 billion pounds to 38.4 billion pounds in 1990 and to 47.4 billion pounds in 1995 — a 45.8 percent increase over the 1984 estimated level. Postconsumer wastes are projected to be much larger than manufacturing waste — accounting for 91 percent of the total in 1984 and 1990 and 92 percent in 1995. Plastic wastes from all postconsumer product categories are projected to rise during this time period, with the packaging and building and construction categories projected to have the largest increases.

While these results are in themselves interesting and important, these estimates and projections are possibly more important in that they suggest which segments of the plastic waste stream will be applicable to different

FIGURE 4.14. **1995 Postconsumer Waste Projections (percentage by plastic type)**

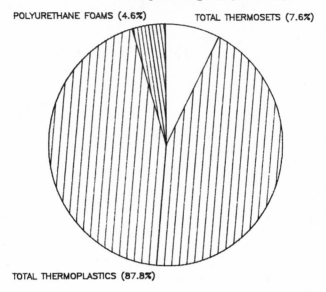

POLYURETHANE FOAMS (4.6%)   TOTAL THERMOSETS (7.6%)

TOTAL THERMOPLASTICS (87.8%)

recycling processes. Recall that in order for secondary recycling to be a viable technological option, plastic wastes must be collected separately from other waste materials. It is generally agreed that once plastics enter the municipal waste stream or are significantly contaminated with other waste materials, secondary recycling is not a viable option and some tertiary recycling process may not be applicable.

If we assume that plastic wastes cannot be separated from other wastes in the municipal waste stream once they enter that stream, there are only three streams of plastic wastes that are suitable for recycling processes that require relatively uncontaminated plastic waste: (a) manufacturing nuisance plastics that can be collected separately from other waste material; (b) postconsumer plastics that by law, regulation, or otherwise are collected separately from other municipal wastes (such as PET bottles collected in states with mandatory deposit laws); and (c) postconsumer plastics that are collected as a by-product of some other recycling operation (such as plastic scrap from automobile recycling). All other plastic wastes will be difficult to divert from the municipal waste stream and therefore can only be disposed of by landfill or incineration or recycled with other municipal wastes in tertiary or quaternary processes.

Unfortunately, plastic wastes from many of the postconsumer product categories will be difficult to divert from the municipal waste stream. Packaging, the largest source of postconsumer plastic waste at about 45 percent of the total, will be very difficult to collect independently of other solid wastes. Packaging is composed almost entirely of thermoplastics, which have characteristics making them easy to recycle in secondary processes. However, packaging is usually bulky, making it difficult to store and transport for the consumer. The only significant segment of packaging waste that is expected to be collected outside of the municipal waste stream is plastic containers that are returned because of bottle deposits. In 1980 plastic bottle production consumed about 1.6 billion pounds of resin, with 66 percent of this total being high-density polyethylene. Other resins used in bottle manufacture were PET at 18 percent, PVC at 6 percent, low-density polyethylene at 5 percent, polypropylene at 3 percent, and polystyrene and other resins at 2 percent.[7] Recall that total packaging waste is estimated to have totaled about 12.5 billion pounds in 1984. Therefore, while waste from plastic bottles may be large in absolute terms, bottles represent a relatively small percentage of total packaging waste. For the vast majority of plastic waste from packaging, the only realistic recycling options will be those that apply to the municipal waste stream.

FIGURE 4.15.   Total Plastic Waste: 1984, 1990, 1995 (billions of pounds)

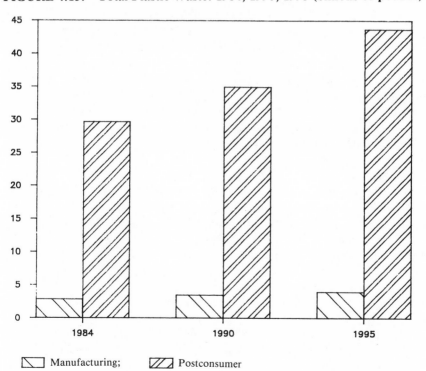

Plastic wastes from consumer and institutional goods face the same problems as those from packaging. This product category is projected to compose more than 10 percent of all postconsumer plastic wastes. Further, the wastes within this category are projected to be mostly thermoplastics at about 95 percent. However, like packaging, it is difficult to prevent these goods from entering the municipal waste stream. The recycling opportunities for this category are therefore the same as those for packaging.

The furniture and the building and construction categories when combined are projected to account for about 16.1 percent of all postconsumer wastes in 1995. However, here again the plastics will be difficult to divert from a general solid waste stream. In both product categories plastics are often used in combination with other materials, which results in a contamination problem. Further, neither furniture nor building and construction products are usually recycled to retrieve their nonplastic materials. Plastics from these categories will therefore be difficult to recycle as relatively uncontaminated wastes.

The only product categories that appear to have any great potential for providing a relatively uncontaminated flow of plastic wastes are the transportation, the electrical and electronic, and the industrial machinery categories. In all three cases these products are often salvaged for ferrous and nonferrous metals. The "scrap" that remains is often composed of a high percentage of plastics. Further, this scrap is often produced in large volumes and could be collected independently of other solid wastes. Plastic wastes from the transportation sector are expected to account for 5.2 percent of total postconsumer waste in 1995, wastes from the electrical and electronic sector 6.4 percent, and wastes from industrial machinery 1.0 percent. Plastics from the transportation sector are expected to consist of about 62.9 percent thermoplastics, 23.2 percent thermosets, and 13.9 percent polyurethane foam in 1995. Thermoplastics, thermosets, and polyurethane foams are expected to account for 83.2, 14.2, and 2.7 percent, respectively, in the electric and electronic goods category. Plastics from industrial machinery are expected to consist of 77.6 percent thermoplastics, with the remainder thermosets. Recall that manufacturing nuisance plastics are projected to consist of 82.2 percent thermoplastics, 13.3 percent thermosets, and 4.5 percent polyurethane foam.

Combining manufacturing nuisance plastics with postconsumer plastics from the transportation, electrical and electronic, and industrial machinery categories it can be projected that about 8.4 billion pounds of waste or 21.9 percent of all plastic wastes could potentially be collected independently of other waste materials in 1990. That number increases to 9.4 billion pounds for 1995 or about 20.0 percent of the total. If we assume that plastic bottle production remains at its 1980 level of 1.6 billion pounds and we assume those bottles can be collected independently from other waste materials, the 1990 percentage increases to 26.2 percent. The 1995 percentage increases to 23.4

percent. The remaining 74 to 80 percent of plastic wastes will be difficult to divert from the municipal waste stream and therefore will not likely be suitable for secondary recycling and some tertiary processes. Tertiary and quaternary recycling opportunities will, however, exist for these plastics as part of the municipal waste stream.

NOTES

1. See Appendix B for a more detailed discussion of the methodology used and the results of the regressions. Also given in Appendix B are the results of regressions in which the historical production of all resins and the use of resins in the eight major product categories are regressed against time, a constant, and an appropriate industrial production index. For the interested reader, these regressions indicate the historical relationships between plastic resin production and use and production trends in several major product categories.

2. See Chapter 6 for a more detailed discussion of the current and projected use of plastics in automobiles. Many experts project that plastics will compose a larger percentage of the future automobile's weight, as manufacturers attempt to improve fuel efficiency by using light-weight plastics. However, recent trends in the use of plastics in automobiles and in the production index for transportation equipment indicate that the use of plastics in domestic transportation equipment will not increase in the coming decade.

3. These projections include only domestic products with plastic content. Unfortunately, detailed information was not obtained that identified the exports and imports of plastic resins by product category. The Society of the Plastics Industry's (SPI) *Facts and Figures of the U.S. Plastics Industry* (1984, p. 10) does, however, provide estimates of the total apparent domestic consumption of selected plastic resins, taking into consideration imports and exports. According to the SPI, this measure is an accurate indication of the actual quantity of resin material consumed in the United States. In 1983 the apparent consumption of several selected resins was about 34.0 billion pounds, while domestic production was about 37.1 billion pounds. Exports totaled 3.8 billion pounds, exceeding the resin import level of 0.6 billion pounds. Therefore, in 1983 apparent consumption of the selected resins was about 91 percent of U.S. production. This compares to about 87 percent in 1980 and about 92 percent in 1976. In some cases the adopted methodology may bias the waste estimates upward. However, in other product categories, such as transportation where imported automobiles have accounted for between 25 and 30 percent of domestic sales in recent years, the projections may be biased downward. Projections given below should be interpreted as rough estimates and not exact levels of future postconsumer plastic wastes.

4. See, for example, Leidner (1981, p. 81) and Milgrom (1973, p. 185) for estimates of the average life times of products in several major product categories.

5. More detailed numerical data can be found in Appendix B. Tables B.9, B.10, and B.11 give numerical results for postconsumer waste estimates for 1984, projections for 1990, and projections for 1995, respectively. In all cases, the data are disaggregated by several specific thermoplastic and thermosetting resins.

6. The "other" category includes the adhesives product category.

7. High-density polyethylene is commonly used for milk bottles and other containers that require a good moisture barrier. PET is commonly used for soft drink containers because it resists permeation of oxygen, water vapor, and carbon dioxide. For more information on the uses of specific resins in bottles see Plastics Beverage Container Division (undated). For more on the recycling of PET bottles see Chapter 6.

5

The Cost of Recycling Versus Disposal

INTRODUCTION

Chapter 3 contains a discussion of the numerous parameters that jointly determine the economic viability of recycling plastic wastes. As discussed in that chapter, both the private and public sectors may consider several potential costs and benefits when making a decision about recycling. For example, the private sector may consider the potential goodwill generated by recycling operations. And, of course, the public sector must consider the potential environmental damage that may be caused by recycling or disposal. However, while these types of considerations are crucial to public and private decision makers, the associated costs and benefits are difficult to quantify. One of the main conclusions of Chapter 3 is that economic viability is not a straight-forward concept. An assessment of viability goes beyond comparing the expected accounting costs and revenues to be generated from a recycling operation with the accounting costs associated with disposal.

This chapter abstracts from the numerous factors that may influence the economic viability of plastics recycling and focuses on published estimates of the expected accounting costs and revenues associated with different recycling and disposal technologies. While a comparison of this type does not allow any definitive conclusions about the economic viability of recycling, it does offer a starting point from which more in-depth comparisons can be made.

Data on the costs and revenues associated with different recycling and disposal technologies were obtained from various sources. All major technical possibilities are represented—i.e., disposal by landfill and incineration without heat recovery, secondary recycling, tertiary recycling, and quaternary recycling. The data unfortunately reflect different time periods. To allow limited comparisons of the data, all costs and revenues have been adjusted to 1981 dollars. In most cases the cost estimates are disaggregated into operating and capital costs. To avoid the problem of dealing with different rates of capital depreciation, capital costs are dealt with as though all capital were rented. In other words, it is assumed that capital investments must earn some real annual gross rate of return. There is no general agreement on what this rate of return should be. Therefore, a rate of 15 percent per year is arbitrarily adopted as a base case. A low rate of 10 percent and a high rate of 20 percent are used as alternative rates to test the sensitivity of the base-case net-cost or net-revenue estimates. In those cases where costs and revenues are disaggreted, the disaggregated parts are adjusted to 1981 dollars using appropriate price indexes. In those cases where the costs and revenues are not disaggregated, the "bottom line" figures (usually a cost per ton) are adjusted to 1981 dollars using the industrial commodities producer price index.

The numbers reported in this chapter are not therefore necessarily those reported in the source documents. They have been altered to reflect constant dollars. Further, in those cases where capital costs are disaggregated from operating costs, adjustments have been made so that capital costs are treated consistently across all estimates. In addition, in several cases adjustments have been made so that revenues from recycled products and costs of residue disposal are treated consistently from estimate to estimate. The "bottom-line" assessments of the costs and revenues reported in this chapter should not therefore be assumed to be consistent with the "bottom-line" assessments reached in the source documents. The reader is encouraged to consult the source documents for information on the bottom-line assessments reached by the authors of those documents.

The following additional caveats are appropriate. Unfortunately, most of the recycling technologies discussed in this chapter have not been used extensively or long enough to provide reliable cost and revenue estimates. Further, many of the estimates have been provided by the developers of the processes. They do not necessarily represent data that have been gathered from users of the processes. Finally, many estimates are somewhat dated or incorporate restrictive assumptions that make comparisons with other technologies difficult. While efforts have been made to allow limited comparisons of the data, the information contained herein should be considered broad cost ranges and not firm cost estimates.[1]

DISPOSAL

Landfill

The cost of landfill varies significantly depending on population size, the distance the waste has to be transported, and the size of the operation. Cost estimates were obtained from several different sources. However, the estimates were not disaggregated into specific operating and capital costs.

The most recent estimates were obtained from Johnson (1985). That source reports the results of a 1984 survey of landfill costs at 87 landfill sites across the United States. The average fee was found to be $10.59 per ton in 1984 dollars or, when deflated by the industrial commodities producer price index, $9.98 in 1981 dollars. In 1981 dollars the cost of landfill ranged from $1.41 to $29.22 per ton.

Table 5.1 presents several more dated landfill cost estimates disaggregated by type and size of operation. In 1981 dollars these estimates range from $1.43 to $15.62 per ton. Table 5.2 presents landfill cost estimates by population size. As one might expect, the cost of landfill generally increases with population. Small communities of less than 10 thousand persons have average landfill costs of $6.98. The average cost of landfill in centers with population of 250–500 thousand is $15.33. The average for all population areas according to this source is $8.23 per ton.

The Metal Scrap Research and Education Foundation (1983) reports the results of an independent informal survey of the landfill cost of several automobile shredder operations. That survey showed a range of cost from $3.65 to $7.79 per ton in 1981 dollars.

Incineration without Heat Recovery

The cost of incineration without heat recovery is generally higher than landfill. The extimates presented in Table 5.3 are between $8.66 and $22.77 per ton including the cost of residue disposal by landfill. The estimates (with one exception) are not disaggregated into operating and capital costs.

SECONDARY RECYCLING

There are numerous published sources of information on secondary recycling processes. (See Chapter 2 for details.) Generally, these secondary processes are designed to recycle one of four types of plastic waste: a) postconsumer plastic waste obtained from returnable bottles; b) mixed manufacturing plastic waste; c) manufacturing plastic waste of a single resin; and d) mixed plastic waste that is collected as part of another recycling operation, such as the residue from automobile shredder operations. Unfortunately, while there are numerous sources for general background information and detailed technical descriptions, there is very little cost information on the proposed and existing secondary recycling technologies. This section reviews four examples identified by the author: two examples of technologies designed to recycle PET bottles; the Mitsubishi Reverzer, which is appropriate for the recycling of mixed manufacturing and postconsumer wastes; and a developmental process to recover polymeric materials from automobile shredder residue.

TABLE 5.1. Cost of Landfill by Type and Size of Operation (all costs in 1981 dollars)

Type of Operation	Tons/Day	Cost/Ton[a]	Source[c]
Close-in	250	$ 6.78	1
Close-in	2,000	5.82	1
Close-in	Not Given	7.34	4
Remote	250	15.09	1
Remote	2,000	13.42	1
Remote	Not Given	15.62	4
Not Given	200	4.30	3
Not Given	500	2.72	3
Not Given	900	1.43	3
Not Given	500	4.24–7.07[b]	2

[a]All costs have been inflated by the industrial commodities producer price index.

[b]Excludes the cost of land.

[c]*Sources*: (1) Huang and Dalton (1975); (2) Diaz, Savage, and Golueke (1982); (3) Goddard (1969); (4) Baum and Parker (1974).

TABLE 5.2. Cost of Landfill by Population Size (all costs in 1981 dollars)

Population (in thousands)	Cost/Ton (1981 $)
>500	13.54
250–500	15.33
100–250	11.80
50–100	7.59
25–50	5.61
10–25	8.32
2.5–10	6.98

Note: All costs have been inflated by the industrial commodities producer price index.

Source: Environmental Protection Agency (1984).

TABLE 5.3. Cost of Incineration without Heat Recovery (all costs in 1981 dollars)

Tons/Day	Cost/Ton[a]	Source[c]
Not Given	$22.77	1
150	20.50	2
200	16.52	2
410	8.66	2
500	22.63[b]	3

[a]All costs have been inflated by the industrial commodities producer price index.

[b]Interest on capital assumed to be 15%. At a 10% interest rate the cost per ton is reduced to $16.71 and at 20% the cost is increased to $28.55 per ton.

[c]*Sources*: (1) Baum and Parker (1974); (2) Goddard (1969); (3) Huang and Dalton (1975).

None of the estimates in this section includes the cost of separating waste plastics from other municipal wastes. Therefore these estimates do not pertain to the cost of secondary recycling of plastic wastes in the municipal waste stream.

PET Recycling

The recycling of postconsumer plastic waste obtained from returnable beverage bottles is becoming increasingly popular, especially in states with mandatory bottle deposit laws. Recycling polyethylene terephthalate (PET) is one of the most popular. The Society of the Plastics Industry (1985) gives information on 39 recyclers and brokers of PET recycled materials in the United States. Specific cost and revenue estimates are often proprietary. However, through informal conversations with several operators it was determined that waste PET is usually purchased in the form of bundles of plastic bottles for about 3–5 cents per pound. After grinding the bottles and removing the paper, metal caps, and (in some cases) the bases of the bottles, which are made from other plastic materials, the relatively pure PET can be sold for about 20 to 30 cents per pound. (If the HDPE material used for the base of the bottle is not recyclable, then approximately 30 percent of the weight of the bottle is waste.) The reclaimed PET can then be applied to several purposes, including cargo straps and stuffing for pillows, ski jackets, and sleeping bags. Reclaimed PET has also been used experimentally to make products such as toys, pails, plastic drums, and sail boats. The recycled PET cannot typically be used for food packaging because of residual traces of the liquid originally contained in the bottles.

One estimate of the cost of recycling PET or other polyester bottles comes from the Goodyear Tire and Rubber Company. While Goodyear is not active in recycling polyester bottles, the company is a major producer of PET and has demonstrated the recycling potential for PET. The cost information provided in Goodyear (undated) was used in combination with other information from currently active PET recyclers and other sources to estimate the costs and revenues from a typical operation. That information is given in Table 5.4. It was assumed that the residue from the recycling process—

TABLE 5.4. Goodyear's Polyester Bottle Recycling Operation (in 1981 dollars)

Capacity per year 5,000 tons	Interest rate (percent)		
Operational hours per year 6,000			
Capital Cost $667,000	10	15	20
Interest on Capital	$ 66,700	$ 100,050	$ 133,400
Miscellaneous			
Operational Cost per year	800,342	800,342	800,342
Material Purchased @5¢/lb	800,000	800,000	800,000
Packaging Cost @1.5¢/lb	105,000	105,000	105,000
Waste Disposal Cost (30% of volume in Landfill @$10.00/ton)	15,000	15,000	15,000
Total Cost	1,487,042	1,520,392	1,553,742
Revenues 7,000,000 lbs.			
@20¢/lb.	1,400,000	1,400,000	1,400,000
@30¢/lb.	2,100,000	2,100,000	2,100,000
Net Revenues or Costs			
@20¢/lb.	– 187,042	– 120,392	– 153,742
@30¢lb.	612,958	579,608	546,258
Net Profit or Loss Per Ton			
@20¢/lb.	– 17.41	– 24.08	– 30.75
@30¢/lb.	122.59	115.92	109.25
Breakeven Cost Per Pound	0.212/lb	0.217/lb	0.222/lb

Sources: Cost data other than landfill and material cost: Goodyear Tire and Rubber Company (undated) (assumed to be in 1981 dollars); landfill cost: Johnson (1985) (deflated to 1981 dollars using the industrial commodities producer price index); prices of processed PET: Informal conversations with PET recyclers; material cost: Informal conversations with PET recyclers.

about 30 percent of the weight of the incoming bottles — is landfilled at a cost of $10 per ton. This reflects the deflated 1981 average U.S. landfill cost given in Johnson (1985). The assumed price range for processed PET reflects information gathered from informal conversations with PET recyclers. It was further assumed that the cost information from Goodyear (undated) reflects 1981 dollars.

Note that the Goodyear process is expected to be profitable at all assumed interest rates, given a market price for the reclaimed PET of 30 cents per pound. At 20 cents per pound, revenues do not exceed expected total costs. Net losses are estimated to be as high as about $31 per ton and net profits as high as about $123 dollars per ton, depending on the particular assumptions. The breakeven price for reclaimed PET ranges from 21.2 to 22.2 cents per pound, depending on the assumed interest rate.

Another estimate of the costs of recycling PET bottles is given in Brown (1979). The description of the process reviewed in Brown (1979) is similar to that given in Goodyear (undated), although the size of the operation is much smaller than the Goodyear process — 1,000 tons per year as compared to 5,000 tons per year. Information on the cost of the operation given in Brown's article was combined with information from other sources to arrive at the estimates presented in Table 5.5. It is assumed, as is the case with the Goodyear process, that 30 percent of the weight of the incoming bottles becomes a waste that must be disposed by landfill at the national average cost of $10 per ton. It is further assumed that incoming material can be obtained at a cost of five cents per pound.

The calculations presented in Table 5.5 indicate that this alternative PET recycling technology is profitable at all assumed interest rates and prices for finished product. Expected profits per ton range from a low of $3.69 to a high of $158.34. The breakeven price for recycled PET in this example is between about 13 and 14 cents per pound, depending on the assumed interest rate.

The Mitsubishi Reverzer

There are several secondary recycling processes that can utilize mixed plastic wastes to produce bulky products that usually compete with wood or concrete. These processes usually involve the melting of the thermoplastic portion of the waste and then cooling the mixture in molds. Products that have successfully been produced using these methods include fence posts, drainage gutters, drain pipes, cable reels, cargo skids, and so forth. These products are typically of as good quality as the products they replace and often have superior qualities because of the plastic's nonbiodegradability. (See Chapter 2 for more technical details.)

Unfortunately, there is very little cost information about these processes.

TABLE 5.5. An Alternative PET Recycling Operation (in 1981 dollars)[a]

| Capacity per year | 1,000 tons | Interest rate (percent) | | |
Capital Cost[b]	$146,494	10	15	20
Interest on Capital		$ 14,649	$ 21,974	$ 29,299
Operating Costs		102,867	102,867	102,867
Material Purchased @$0.05/pound		100,000	100,000	100,000
Waste Disposal Cost (30% of volume in landfill @$10.00/ton)		3,000	3,000	3,000
Miscellaneous costs		41,147	41,147	41,147
Total Costs		261,663	268,988	276,313
Revenues 1,400,000 pounds				
@$0.20/pound		280,000	280,000	280,000
@$0.30/pound		420,000	420,000	420,000
Net Revenues or Costs				
@$0.20/pound		18,337	11,012	3,687
@$0.30/pound		158,337	151,012	143,687
Net Profit or Loss Per Ton				
@$0.20/pound		18.34	11.01	3.69
@$0.30/pound		158.34	151.01	143.69
Breakeven Cost Per Pound		0.131	0.135	0.138

[a]Inflated by industrial commodities price index unless otherwise noted.
[b]Inflated by capital equipment price index.
Sources: Cost of capital, operating cost, and miscellaneous cost: Brown (1979); landfill cost: Johnson (1985); price of processed PET: Informal conversations with PET recyclers.

The sole exception identified by this author is a set of cost and revenue estimates given in a 1974 article in *European Plastics News*. That article describes one of the most popular secondary recycling processes – the Mitsubishi Reverzer – and gives cost and revenue estimates for the production of fence posts and cable reels. The process is suitable for making several heavyweight products as long as the waste is composed of between 50 and 80 percent thermoplastics. (The *European Plastics News* article reports that the process requires the waste to be composed of at least 80 percent thermoplastics. However, other sources, discussed in Chapter 2 of this book, report that the process can handle waste that consists of as little as 50 percent thermoplastics.) It is assumed that the required waste material can be obtained at a cost of 5 cents per pound for transportation and handling (as is assumed with the PET recycling processes).

According to the calculations presented in Tables 5.6 and 5.7, the Rever-

TABLE 5.6. The Mitsubishi "Reverzer" (fenceposts) (in 1981 dollars)[a]

Capacity per year Capital Cost[b]	1,240 tons $753,308	Interest rate (percent)		
		10	15	20
Interest on Capital		$ 75,330	$112,996	$150,662
Operating Costs				
1,240 Tons of Waste @5¢/lb.		124,000	124,000	124,000
Electricity[c]		30,030	30,030	30,030
Cooling Water[d]		1,845	1,845	1,845
Labor[d]		206,731	206,731	206,731
Miscellaneous[d]		114,081	114,081	114,081
Total Operating Cost		476,687	476,687	476,687
Total Cost		552,017	589,683	627,349
Revenues				
300,000 Fence Posts per year @$2.56 each[e]		768,000	768,000	768,000
Net Revenues		215,983	178,317	140,651
Profit per ton		174.18	143.80	113.42

[a]Inflated by industrial commodities price index unless otherwise noted.
[b]Inflated by capital equipment price index.
[c]Reflects average cost of electricity to industrial users in 1981: 4.29¢ per kilowatt-hour (see any recent issue of the U.S. Department of Energy's *Monthly Energy Review*).
[d]Adjusted to 1981 dollars using the price index for all commodities.
[e]Inflated by lumber and wood products price index.
Note: It is assumed that the cost of obtaining material is equivalent to the transportation and handling cost associated with waste PET.
Source: estimated operating cost data (exclusive of cost of obtaining waste material), capital cost data, and revenue data: *European Plastics News* (1974).

zer is a very profitable process under all assumed costs of capital. These estimates should, however, be interpreted cautiously. The estimates are based on experience with the technology in a country outside the U.S. and are from the 1973 time period. While attempts have been made to adjust the data so limited comparison with other estimates can be made, these characteristics of the data may pose problems.

The Secondary Recycling of Auto Shredder Residue

Work is currently being sponsored by the Energy Conversion and Utilization Technologies Program within the U.S. Department of Energy (DOE) to develop processes to utilize the residue from automobile shredder operations. This residue can contain relatively high percentages of plastics—as high as about 36 percent according to one estimate. (See Chapter 6 for a detailed discussion of the potential recycling of plastics from automobile shredder residue.) As part of the DOE program, cost estimates have recently been completed for the processing of the waste into a compounded polymeric material, which would be suitable for forming into various products. [See Plastics Institute of America (draft) for details.] The process would involve the separation and sale of ferrous and nonferrous metals from the residue. A binding material would be mixed with the plastics portion, and it is assumed by the study that the remaining residue portion would be sold as clean fill.

Table 5.8 presents the results of six selected modes of operation. The study by the Plastics Institute of America considered three alternative prices for the compounded polymeric material: 10, 12.5, and 15 cents per pound. Table 5.8 incorporates the midrange price adjusted for sales returns and allowances and sales discounts. It is assumed that the waste material could be obtained at zero cost. It is further assumed that the operation would be

TABLE 5.7. The Mitsubishi "Reverzer" (cable reels) (in 1981 dollars)[a]

Capacity per year Capital Cost[b]	1,047 tons $719,067	Interest rate (percent)		
		10	15	20
Interest on Capital		$ 71,907	$107,860	$143,813
Operating Costs				
1,047 Tons @5¢/lb.		104,700	107,700	104,700
Electricity[c]		30,030	30,030	30,030
Cooling Water[d]		1,845	1,845	1,845
Labor[d]		206,731	206,731	206,731
Miscellaneous[d]		110,657	110,657	110,657
Total Operating Cost		453,963	453,963	453,963
Total Cost		525,870	561,823	597,776
Revenues				
36,000 Wood Replacement Reels @$19.26 each[e]		693,360	693,360	693,360
Net Revenues		167,490	131,537	105,584
Profit per ton		159.97	125.63	100.84

[a]Inflated by industrial commodities producer price index unless otherwise noted.
[b]Inflated by capital equipment price index.
[c]Reflects average cost of electricity to industrial users in 1981; 4.29¢ per kilowatt-hour.
[d]Adjusted to 1981 dollars using the price index for all commodities.
[e]Inflated by lumber and wood products price index.
Note: It is assumed that the cost of obtaining material is equivalent to the transportation and handling cost associated with waste PET.
Source: estimated operating cost data (exclusive of cost of obtaining waste material), capital cost data, and revenue data: *European Plastics News* (1974).

TABLE 5.8. The Secondary Recycling of Auto Shredder Residue (in 1981 dollars)[a]

	2-Shift Operation	3-Shift Operation	2-Shift Operation	3-Shift Operation	2-Shift Operation	3-Shift Operation
Capacity (tons/yr) (approximate)	3,864	3,864	11,591	11,591	19,318	19,318
Capital Cost[b]	1,179,510	943,608	1,651,313	1,415,411	2,359,019	1,887,215
Interest @10%	117,951	94,361	165,131	141,541	235,901	188,722
Interest @15%	176,927	141,541	247,697	212,312	353,853	283,082
Interest @20%	235,902	188,722	330,263	283,082	471,804	377,443
Operating Cost	712,359	775,517	1,556,592	1,642,374	2,400,825	2,509,230
Revenues						
CPM Sales[c] @12.5¢/lb. (1984 dollars) (less sales returned and allowances and sales discounts)	544,448	544,48	1,633,343	1,633,343	2,722,239	2,722,239
By-Product Sales						
Clean Fill	19,121	19,121	57,362	57,362	95,604	95,604

Ferrous Metal[d]	45,911	45,911	27,546	27,546	9,182
Nonferrous Metal[d]	458,841	458,841	275,468	275,468	91,823
Total Revenues	3,322,595	3,322,595	1,993,719	1,993,719	664,574
Net Revenues or Costs					
Interest @10%	624,643	685,869	209,804	271,996	−165,736
Interest @15%	530,283	567,917	139,033	189,430	−224,712
Interest @20%	435,922	449,966	68,263	106,864	−283,687
Profit or Loss Per Ton					
Interest @10%	32.33	35.50	18.10	23.47	−42.90
Interest @15%	27.45	29.40	12.00	16.34	−58.16
Interest @20%	22.57	23.29	5.89	9.22	−73.42

[a]Deflated by industrial commodities price index unless otherwise noted.
[b]Capital Cost deflated by capital equipment price index.
[c]Compounded Polymeric Material (CPM) deflated by rubber and plastic products price index.
[d]Deflated by metals and metal products price index.
Source: data before adjustment to 1981 dollars: Plastics Institute of America (draft).

103

installed at the shredder site thus negating the need for transportation of the waste material. Recall that in the previous two secondary recycling examples, a 5 cent per pound transportation and handling charge was assessed. All costs and revenues presented in Table 5.8 have been deflated to 1981 dollars using the price indexes indicated in the table.

Note that the profitability of the developmental process varies greatly depending on operation size. At about 3,900 tons per year the process results in expected losses of over $73 per ton at an assumed 20 percent interest rate. However, at a capacity level of over 19,000 tons per year the process is expected to produce profits of between $22 and $32 dollars per ton.

TERTIARY RECYCLING

Recall that tertiary recycling refers to any process that converts a waste into a basic chemical or fuel product. The most common of these processes is pyrolysis, and it is this process on which this section focuses.

When discussing the pyrolysis of plastic wastes, we must consider the process at two levels: a) plastics as part of the municipal waste stream; and b) pyrolysis of relatively uncontaminated plastic wastes. Cost and revenue estimates for municipal pyrolysis systems are fairly well documented. Estimates for the pyrolysis of less contaminated wastes are more limited.

Pyrolysis of Municipal Waste

Huang and Dalton (1975) provide detailed cost and revenue estimates for several commercially available waste pyrolysis systems. The basic data provided by Huang and Dalton have been inflated to reflect 1981 dollars and have been adjusted so that cost-per-ton estimates will be consistent with other estimates presented in this chapter. (See the introduction to this chapter for details.) It is assumed that the waste to be processed can be obtained at zero monetary costs. Operating costs include the cost of disposing residue from the pyrolysis operations by landfill. In some cases note that certain materials, such as metals and glass, are separated from the waste stream and sold separately.

Results are presented in Table 5.9. Note that not all processes discussed in Huang and Dalton (1975) are included in Table 5.9. The processes included are representative of the quite extensive information presented by Huang and Dalton. It is suggested that the interested reader refer to that report for a discussion of the technical characteristics of the particular technologies and further details about the economic assessments.

The per ton estimates range from a profit of over $23 per ton to a loss of over $19 per ton. Moreover, two estimates from different sources of the same process vary significantly. The Garrett 2000 tons per day operation at one site has an expected net cost of $11.15 per ton, while a similar process reported by another source has an expected net profit of $11.34 per ton at an assumed 15 percent interest rate.

Table 5.10 presents estimates of the net costs of municipal waste pyrolysis from sources other than Huang and Dalton (1975). The estimates given in Table 5.10 could not be disaggregated into various cost and revenue components. The "bottom-line" estimates given in the cited reports have therefore been adjusted to 1981 dollars using the industrial commodities producer price index. As with Table 5.9, the interested reader is referred to the source documents for information on the specific technical characteristics of the different pyrolysis processes, as well as more detailed discussions of the economic analyses relating to each technology. As is the case with most other tables in this chapter, the estimates in Table 5.10 vary widely—from as little as $1.91 per ton to as high as $27.00 per ton.

Pyrolysis of Plastics

Two references were identified that document the cost of a pyrolysis process designed to convert relatively clean thermoplastic wastes into gaseous fuels, No.6 fuel oil, and No.2 fuel oil. The process has been used by U.S.S. Chemicals and is discussed in *Chemical Engineering* (1982) and in Machacek (1983). (Note that Machacek's assessment does not include the production of gaseous fuels.) Also see Chapter 2 of this book for additional discussion of the technology. While the technology can process a wide variety of wastes, polyethylene scrap from resin producers, fabricators, and converters is particularly applicable.

Table 5.11 presents cost and revenue estimates for the process, which have been derived, in part, from information given in *Chemical Engineering* (1982). Table 5.12 presents an assessment of the U.S.S. Chemicals process based, in part, on information from Machacek (1983). To be consistent with other processes that require relatively clean waste, it is assumed in both cases that the input material can be obtained at a cost of five cents per pound. The calculations presented in Table 5.11 suggest that the process is profitable at all interest rates considered—ranging from $11.15 to $35.63 per ton. Results presented in Table 5.12 indicate that the profitability of the process varies significantly with the interest rate, ranging from a profit of $16.81 per ton to a loss of $11.73.

QUATERNARY RECYCLING

If plastic wastes cannot be recycled in a secondary or tertiary sense, they can be burned to retrieve at least part of their heat energy. As was the case with pyrolysis, quaternary recycling must be considered at two main levels: a)

TABLE 5.9. **Municipal Waste Pyrolysis (in 1981 dollars)**

	Process							
	Union Carbide	Union Carbide	Garrett	Garrett	Garrett	Garrett	Garrett	West Virginia University
Capacity (tons/day)	2,000	1,000	2,000	2,000	250	500	1,000	1,000
Operational days/year	365	365	365	300	350	350	350	320
Capital Cost[a]	$52,217,975	$22,971,626	$59,922,267	$45,369,716	$8,988,340	$15,408,583	$26,322,995	$25,038,947
Interest @10%	5,221,798	2,297,163	5,992,226	4,536,972	898,834	1,540,858	2,632,300	2,503,895
Interest @15%	7,832,696	3,445,744	8,988,340	6,805,457	1,348,251	2,311,287	3,948,449	3,755,842
Interest @20%	10,443,595	4,594,325	11,984,453	9,073,943	1,797,668	3,081,717	5,264,599	5,007,789
Operating Cost[b]	6,009,643	2,678,295	13,696,036	6,168,753	1,212,708	2,666,297	7,176,575	5,426,731
Revenues								
Liquid Fuels[c]			12,918,328	18,073,079	1,613,472	3,226,945	6,453,891	9,063,472
Gas Fuels[d]	27,486,142	7,530,449						
Solid Fuels[e]								
Steam Power[f]								
Electricity[g]								
Magnetic Metals[h]	728,284	448,942	1,152,286	1,002,638	125,329	250,659	501,319	
Nonmagnetic Metals[i]								

Glass^j Paper^k			472,086				462,451	
Total Revenues	28,214,425	7,979,391	14,542,700	20,000,620	1,854,413	3,708,829	7,417,661	9,063,472
Net Revenue or Cost								
Interest 10%	16,982,985	3,003,933	−5,145,562	9,294,895	−257,129	−498,326	−2,391,214	1,132,846
Interest 15%	14,372,086	1,855,352	−8,141,676	6,805,451	−706,546	−1,269,755	−3,707,363	−119,101
Interest 20%	11,761,187	706,771	−11,137,789	4,757,924	−1,155,963	−2,039,185	−6,935,488	1,371,048
Profit or Cost per Ton								
Interest 10%	23.26	8.23	−7.05	15.49	−2.94	−2.85	−6.83	3.54
Interest 15%	19.69	5.08	−11.15	11.34	−8.07	−7.25	−10.59	−.37
Interest 20%	16.11	1.94	−15.26	7.93	−13.21	−11.65	−19.82	−4.28

[a]Inflated by capital equipment price index.
[b]Inflated by GNP implicit price index.
[c]Inflated by producer price index for refined petroleum products.
[d]Inflated by producer price index for gas fuels.
[e]Inflated by producer price index for coal.
[f]Inflated by producer price index for fuels and related products.
[g]Inflated by producer price index for electric power.
[h]Inflated by producer price index for iron and steel.
[i]Inflated by producer price index for nonferrous metals.
[j]Inflated by producer price index for nonmetallic minerals.
[k]Inflated by producer price index for pulp, paper, and allied products.

Source: All data prior to adjustment to 1981 dollars from Huang and Dalton (1975).

TABLE 5.10. Additional Estimates for Municipal Waste Pyrolysis (in 1981 dollars)

Type of Operation	Capacity (tons/day)	Net Cost/Ton	Source
Union Carbide (PUROX)	200	$1.91 to $4.19	1
PUROX	1,000	$23.15 to $25.72	2
TORRAX	1,000	$15.43 to $27.00	2
Not Given	Not Given	$19.37	3
Landgard	1,000	$9.71	4

Note: All costs have been inflated to 1981 dollars using the industrial commodities producer price index.

Sources: (1) Schulz (1975); (2) Office of Technology Assessment (1979); (3) Baum and Parker (1974); (4) *Environmental Science and Technology* (1975).

TABLE 5.11. U.S.S. Chemicals Plastic Waste Pyrolysis Process (cost information as reported in *Chemical Engineering*, 1982) (in 1981 dollars)

Capacity (ton/year) Capital Cost	12,500 $3,060,000	Interest Rate (percent)		
		10	15	20
Interest on Capital		$ 306,000	$ 459,000	$ 612,000
Operating Cost		496,250	496,250	496,250
Plastic Waste Acquisition Cost[a]				
(5¢/lb.)		1,250,000	1,250,000	1,250,000
Total Cost		2,052,250	2,205,250	2,358,250
Revenues				
No.6 Fuel Oil[b]		252,641	252,641	252,641
No.2 Fuel Oil[c]		2,140,612	2,140,612	2,140,612
Gaseous Fuel[d]		104,330	104,330	104,330
Total Revenues		2,497,583	2,497,583	2,497,583
Net Revenues		445,333	292,333	139,333
Profit per ton		35.63	23.38	11.15

[a]It is assumed, as with other recycle processes that require relatively pure plastic waste, that wastes of this type cost 5 cents per pound to obtain at the recycle location.

[b]6.9 million lbs./year @0.12687 gal./lb. @$0.2886/gal.

[c]15.1 million lbs./year @0.13817 gal./lb. @$1.026/gal.

[d]1.0 million lbs./year @0.22104 gal./lb. @0.472/gal (the wholesale price of propane in 1981).

Sources: Capital and operating costs and production of levels of fuel oil: *Chemical Engineering* (1982); information used in footnotes b, c, and d: American Petroleum Institute (1984) and U.S. Department of Energy's *Monthly Energy Review* (any recent issue).

TABLE 5.12. U.S.S. Chemicals Plastic Waste Pyrolysis Process (cost information as reported in Machacek, 1983) (in 1981 dollars)[a]

Capacity per year Capital Cost[b]	8,500 tons $2,430,000	Interest Rate (percent)		
		10	15	20
Interest on Capital		$ 243,000	$ 364,500	$ 486,000
Operating Cost		366,600	366,600	366,600
Plastic Waste Acquisition Cost[c] (5¢/lb.)		850,000	850,000	850,000
Total Cost		1,459,600	1,581,100	1,702,600
Revenues				
No.6 Fuel Oil[d]		759,286	759,286	759,286
No.2 Fuel Oil[d]		843,240	843,240	843,240
Total Revenues		1,602,526	1,602,526	1,602,526
Net Revenues		142,926	21,426	− 100,074
Profit or Loss per ton		16.81	2.52	− 11.73

[a]Deflated by the industrial commodities price index unless otherwise noted.
[b]Deflated by the capital equipment price index.
[c]It is assumed, as with other recycling processes that require relatively clean plastic waste, that wastes of this type can be obtained at the recycling location at a cost of 5 cents per pound.
[d]Deflated by the fuels and related products price index.
Source: capital and operating costs and revenues received from fuel oil: Machacek (1983).

incineration with heat recovery of municipal waste, which may contain between 4 and 6 percent plastics; and b) incineration with heat recovery of relatively uncontaminated plastic wastes. Most of the commonly used plastics have heating values similar to coal. (See Table 2.1 for a listing of the Btu contents of various plastic resins and several conventional and unconventional fuel sources.) Further, recall from Chapter 2 that it is generally concluded that the incineration of plastics as part of the municipal waste stream does not pose significant environmental problems.

Incineration of Plastics in Municipal Waste with Heat Recovery

Huang and Dalton (1975) provide several cost and revenue estimates for municipal waste incineration systems with some form of heat recovery. The information given in that source has been inflated to 1981 dollars and adjusted as described in the introduction of this chapter to allow limited comparisons with other estimates. Results are reported in Table 5.13. Note that in some

TABLE 5.13. Incineration of Municipal Waste with Heat Recovery (in 1981 dollars)

			Process					
	Black Clawson	Saugus[a] Waterwall Incinerator	CPU-400	CPU-400 (Without Materials Recovery)	CPU-400	Horner Shifrin	Horner Shifrin	Horner Shifrin
Capacity (tons/day)	2,000	1,200	600	600	1,000	490	980	1,550
Operational Day/Year	300	300	300	300	365	312	312	312
Capital Cost[b]	$77,042,915	$65,094,842	$19,474,736	$17,976,680	$19,915,593	$9,769,464	$11,151,961	$19,902,753
Interest @10%	7,704,292	6,509,484	1,947,474	1,797,668	1,991,559	976,946	1,115,196	1,990,275
Interest @15%	11,556,437	9,764,226	2,921,210	2,696,502	2,987,338	1,465,420	1,672,794	2,985,413
Interest @20%	15,408,583	13,018,968	3,894,947	3,595,336	3,983,119	1,953,893	2,230,392	3,980,551
Operating Cost[c]	11,960,959	6,567,452	1,808,910	1,624,327	2,170,692	900,763	1,655,522	8,730,154
Revenues								
Solid Fuel[d]								
Steam Power[e]		17,111,107				835,314	1,670,629	7,061,805
Electricity[f]	5,342,923		2,255,901	2,255,901	2,989,068			
Hot Water[e]					1,236,365			
Magnetic Metals[g]	1,122,356		848,002		456,425			
Nonmagnetic Metals[h]	1,381,333		657,777		750,277			

Glass[i]	578,064		216,774		315,526		
Paper[j]	4,220,845						
Total Revenues	12,645,521	17,111,107	3,978,454	2,255,901	5,547,661	835,314	1,670,629	7,061,805
Net Revenue or Cost								
Interest @10%	−7,019,730	4,034,171	222,070	−1,166,094	1,385,410	−1,042,395	−1,100,089	−3,658,624
Interest @15%	−10,871,875	779,429	−751,666	−2,064,928	389,631	−1,530,869	−1,657,687	−4,653,762
Interest @20%	−14,724,021	−2,475,313	−1,725,403	−2,963,762	−606,150	−2,019,342	−2,215,285	−5,648,900
Profit or Cost per Ton								
Interest @10%	−11.70	11.21	1.23	−6.48	3.80	−6.82	−3.60	−7.57
Interest @15%	−18.12	2.16	−4.18	−11.47	1.06	−10.01	−5.42	−9.62
Interest @20%	−24.54	−6.86	−9.59	−16.47	−1.66	−13.21	−7.25	−11.68

[a] Assumes that resource recovery just pays for residue disposal cost.
[b] Inflated by Capital Equipment Price Deflator.
[c] Inflated by GNP Implicit Price Deflator.
[d] The solid fuel produced from the Horner Shifrin process competes with coal and was worth about $3.00 per ton at the time of the survey. This price has been inflated by the producer price index for coal.
[e] Inflated by producer price index for fuels and related products.
[f] Inflated by producer price index for electric power.
[g] Inflated by producer price index for iron and steel.
[h] Inflated by producer price index for nonferrous metals.
[i] Inflated by producer price index for nonmetallic minerals.
[j] Inflated by producer price index for pulp, paper, and allied products.
Source: All data prior to adjustment to 1981 dollars from Huang and Dalton (1975).

cases the processes involve the separation and sale of metals, glass, and paper. Once again we find that the estimates differ widely. At a 15 percent interest rate, the results vary from a net cost of $18.12 per ton to a net profit of $2.16.

Table 5.14 presents several estimates of the net cost of incineration of municipal waste with heat recovery from sources other than Huang and Dalton (1975). These estimates could not be disaggregated into their various cost and revenue components. Therefore, all costs per ton have been adjusted to 1981 dollars using the industrial commodities producer price index. See the source documents for discussions of the technical characteristics of the different technologies. Cost estimates vary widely from a low of $2.39 per ton to a high of $30.72 per ton depending on the source and particular process.

The range of estimates presented in Table 5.14 are generally in line with a 1984 survey of the disposal fees at several resource recovery projects in the United States. The results of the survey are reported in Johnson (1985). A total of 13 resource recovery facilities were surveyed that appeared to receive little or no government support. The average 1984 disposal fee at those sites

TABLE 5.14. Additional Estimates of the Cost of Incinerating Municipal Waste with Heat Recovery (in 1981 dollars)

Process	Capacity (tons/day)	Cost per Ton	Source
Horner-Shifrin (Refuse Derived Fuel)	600	$2.39 to $3.93	1
Waterwall Incineration to Steam	1,000	$11.57 to $21.86	2
Refuse Derived Fuel with Materials Recovery	1,000	$5.14 to $12.86	2
Refined Refuse Derived Fuel with Materials Recovery	1,000	$12.86 to $15.43	2
Wet Process Refuse Derived Fuel with Materials Recovery	1,000	$11.57 to $20.57	2
Modular Incineration with Heat Recovery	1,000	$3.86 to $15.43	2
Incineration with Steam and Residue Recovery	Not Given	$22.39	3
Incineration with Steam Recovery	Not Given	$22.94	3
Incineration with Energy Recovery (Electrical Generation)	Not Given	$30.72	3

Note: All costs have been inflated to 1981 dollars using the industrial commodities producer price index.

Sources: (1) Schulz (1975); (2) Office of Technology Assessment (1979); (3) Baum and Parker (1974)

was $17.26 per ton or $16.27 per ton in 1981 dollars (deflated by the industrial commodities producer price index). The range of disposal fees was from $2.83 to $28.29 per ton in 1981 dollars.

Incineration of Relatively Uncontaminated Plastic Waste

One source was identified that gives estimates of the costs and revenues associated with a process that retrieves heat energy from relatively uncontaminated plastic wastes. The process developed by Industronics Incorporated has been used to recycle PET bottles, but can be used to burn other plastics that transform into a liquid prior to complete combustion. Test burns from the rotary kiln device have shown that environmental pollutants are minimal.

Information from Industronics, Inc. (1982) was combined with information on the value of fuel from other sources to estimate costs and revenues from the process. Those results are reported in Table 5.15. It is assumed that the waste material can be obtained at the point of incineration at a cost of 5 cents per pound. Depending on the assumed interest rate, the process is expected to result in either a small profit or loss.

CONCLUSIONS

The main conclusion to be drawn from this chapter is that the current quantity and quality of information about the expected costs and revenues associated with different recycling processes do not justify any definitive conclusions about the competitiveness of recycling with disposal. In some cases the recycling technologies have not been used commercially, and in those cases where the technologies have been commercialized, experience has been limited. The numbers concerning recycling are preliminary, and any assessment of the competitiveness of recycling with disposal should be considered likewise.

However, given these caveats, the currently available numbers suggest some general conclusions. First, in many locations within the United States, recycling of plastics as a relatively uncontaminated waste appears to be competitive with, or superior to, disposal. The cost of landfill averages about $10 per ton in 1981 dollars, with a range of between $1.41 to $29.22 per ton. Population centers of between 2.5 and 10 thousand persons pay on average $6.98 per ton, while areas with populations between 250 and 500 thousand pay an average of $15.33 per ton.[2] Incineration without heat recovery ranges between about $9 and $23 per ton. In most cases, the recycling of relatively clean plastic wastes by secondary, tertiary, or quaternary means will have net costs below these costs of disposal, and in many cases significant profits can be expected. However, given that disposal costs are very low in some locations and given that many recycling programs require large quantities of wastes to

TABLE 5.15. Industronic's PET Incineration Process (in 1981 dollars)

Capacity (ton/year)	4,998	Interest Rate (percent)		
Operational Days/Year	250			
Capital Cost	$597,696	10	15	20
Interest on Capital		$ 59,770	$ 89,654	$119,539
Operating Cost		288,324	288,324	288,324
Plastic Waste				
Acquisition Cost (5¢/lb)		499,800	499,800	499,800
Total Cost		847,894	877,778	907,663
Revenues				
Energy Value[a]		887,732	887,732	887,732
Net Revenues or Cost		39,838	9,954	− 19,931
Profit or Loss per Ton		7.97	1.99	− 3.99

[a]The manufacturer claims the process can produce 20,000,000 Btu/hr. Distillate fuel oil produces 4,825,000 Btu/barrel. Therefore, the incineration process products heat equivalent to 20,600 barrels of distillate fuel oil per year, or 865,236 gallons. The wholesale price of distillate fuel oil in 1981 was $1.026 per gallon.

Sources: Capital and operating costs and Btu output from process: Industronics, Inc. (1982); information used in footnote[a]: American Petroleum Institute (1984) and U.S. Department of Energy's *Monthly Energy Review* (any recent issue).

be economically viable, the recycling of relatively uncontaminated waste plastics will probably not be suitable for all locations.

Second, the recycling of municipal waste, in which plastics are a relatively small part, is generally more costly than disposal. While some estimates were obtained that suggest municipal-waste recycling could be very profitable, the more recent estimates indicate that pyrolysis of municipal waste will produce net costs in the range of about $15 to $27 dollars per ton. Incineration with heat recovery is currently about $16.27 per ton and ranges between about $3 and $28 per ton. These estimates for the tertiary or quaternary recycling of municipal waste are not, however, vastly higher than the cost of disposal. Therefore, in areas with high disposal costs, we can expect continued movement to some form of municipal-waste recycling — recycling that will utilize the plastics component in a tertiary or quaternary sense.

Third, the calculations suggest that plastics that can be easily segregated from other municipal wastes can be recycled more economically outside of, rather than as a part of, the municipal waste stream. We cannot, however, argue that any one form of recycling — secondary, tertiary, or quaternary — is the clear favorite. With the exception of the estimates concerning the

Mitsubishi Reverzer, the cost and revenue calculations do not differ enough to suggest a winner. For example, the secondary recycling of PET bottles is expected to be profitable at between 13 and 22 cents per pound for processed PET, depending on the particular study cited. And clean PET is currently selling in the range of 20 to 30 cents per pound. The analysis of the secondary recycling of automobile shredder residue depends greatly on the assumed capacity. However, profits of over $16 dollars per ton are expected for an operation of about 11.5 thousand tons per year and a product price of 12.5 cents per pound in 1985 dollars. The pyrolysis of relatively clean plastics is expected to produce profits of over $23 per ton, and the incineration of PET is expected to produce profits of about $2 per ton, given similar market assumptions. Given the current quantity and quality of data, we cannot at this time conclude that one form of recycling is necessarily superior to another form.

NOTES

1. Much of the analysis contained in this chapter is taken from a previous publication by this author; see Curlee (1984a).

2. All dollar figures in this section are given in 1981 dollars unless otherwise stated.

6

Incentives and Barriers to Recycling in Selected Sectors

INTRODUCTION

Several general conclusions have been drawn in previous chapters. In Chapter 2 we concluded that there are numerous secondary, tertiary, and quaternary processes that are suitable for the recycling of plastic wastes with varying levels of contamination. Recycled products range from heat energy to sailboats. Chapter 3 was devoted to a conceptual discussion of the economic and institutional incentives and barriers that influence the decision to recycle by the public and private sectors. It was concluded that, while recycling faces significant barriers, there are many incentives that encourage the public sector and the different private-sector actors to promote or adopt the recycling of relatively uncontaminated plastic wastes as an alternative to disposal. In Chapter 4 we concluded that about 75 percent of all plastic wastes in the 1990–95 time period will be difficult to divert from the municipal waste stream. Because the separation of plastics from other similar waste materials is a formidable task, recycling possibilities for these plastics would seem to be limited to tertiary and quaternary recycling as part of the municipal waste stream. However, as much as 25 percent of all plastic wastes have the realistic potential to be collected separately from other municipal wastes. For this significant portion — as much as 11 billion pounds per year — recycling technologies that require relatively clean plastics would be applicable. Finally, in Chapter 5 we concluded that for plastics that can be collected separately from other municipal wastes, recycling outside of the municipal waste stream will in most cases be less costly than either the disposal or the recycling of the plastics as part of the municipal waste stream. Therefore, there is reason to believe that the recycling of plastic wastes is a viable alternative to disposal

116

in many cases. However, evidence to date suggests that recycling technologies have not penetrated their potential markets to a great extent.

This chapter contains a review of three market sectors where secondary, tertiary, and/or quaternary recycling of plastic wastes is currently taking place or where recycling has a good chance of being implemented in future years. The sectors include the electrical and electronics sector (with a particular emphasis on telephones), carbonated beverage containers manufactured from PET, and shredder residue from automobile recycling operations. Although these examples do not represent all the areas where recycling is currently active or where good possibilities for recycling exist, they are a representative sample of the types of waste streams that lend themselves to recycling as relatively uncontaminated waste, rather than as part of the municipal waste stream. They are also among the most discussed examples of recycling found in the literature.

In each example, specific technical, economic, and institutional incentives and barriers are discussed that have a major impact on the recyclability of the waste stream. A major purpose of this chapter is to provide background information on these important market segments. A more basic purpose, however, is to identify the technical, economic, and institutional conditions that have allowed or encouraged the recycling of these waste streams rather than their disposal. This may help identify other market areas where recycling outside of the municipal waste stream is a viable alternative.

Each of the following sections, addressing these three market sectors, is divided into two major parts. The first contains general background information about the recycling operations in that particular sector. The discussions of the technologies in the first part draw from the review of technological issues presented in Chapter 2. The second part is devoted to the specific economic and institutional incentives and barriers that have helped make recycling a viable option to disposal in these selected sectors. The discussion of economic and institutional issues presented in Chapter 3 serves as the basis for these assessments.

THE ELECTRICAL AND ELECTRONICS SECTOR

Background and Technological Issues

There are three main examples of plastics recycling within the electrical and electronics sector: the recycling of ABS (acrylonitrile-butadiene-styrene) from telephone equipment; the recycling of polymeric insulation and sheathing from cable scrap — predominantly PVC (polyvinyl chloride) and polyethylene; and the recycling of epoxy resins from circuit boards. The three examples present an interesting array of technical, economic, and institutional incentives and barriers.

Telephone Equipment

The recycling of ABS from telephone equipment has been developed jointly by Western Electric and Bell Laboratories. ABS, a thermoplastic that possesses outstanding impact and mechanical strength and exhibits good electrical insulating properties, has been used to manufacture telephone bodies since the early 1960s. Thermosetting plastics, such as phenolics, were used prior to that time. Historically, most telephone equipment in the United States has been serviced by the Bell System, and up until the mid 1970s the equipment that was not repairable was usually disposed by landfill. However, in recent years a growing percentage of waste telephone bodies have been recycled in a secondary sense to recover the relatively pure ABS.

The first step in the recycling process is grinding by hammermill. This step is followed by separation using aspirators, screens, and magnets to remove most of the magnetic and nonplastic parts, such as copper, cotton, and labels. Non-ABS plastics are then separated from other plastics in the pulverized scrap by flotation. Additional screening following the flotation step results in a relatively pure ABS scrap.

The finely ground ABS has been used to produce a variety of secondary products, such as rectangular trays for transportation and storage, cable supports, and connector brackets for electrical products. The ABS is not used to manufacture new bodies for telephones because it is difficult to control the color of the recycled resin due to telephones of various colors in the incoming scrap. Another potential use for the ABS is a replacement for the PVC that is currently used to manufacture plastic conduit. While ABS does not have the same flame-retardant characteristics exhibited by PVC, chemical additives can increase the flame retardancy of ABS to an acceptable level. According to Wehrenberg (1982), the recycling operation, other than the flotation separation equipment, consists of commercially available equipment from other industries. That same article reports that about two million pounds of ABS were recycled by Western Electric in 1982, with a potential of as much as ten million pounds that could be recycled each year. It is estimated in Appendix B of this book that a total of about 224 million pounds of ABS and SAN (styrene-acrylonitrile) entered the postconsumer waste stream in 1984 from the electrical and electronics sector.[1] (ABS is an impact-modified version of SAN.)

Electric Cable

Significant progress has also been made toward the recycling of thermoplastics from electric cable insulation and sheathing. The recycling of copper cable has been the norm for a number of years. However, until recently the plastics used for insulation and sheathing, usually PVC and polyethylene, were disposed by landfill. Processes have now been developed to recycle this relatively uncontaminated source of thermoplastics into various secondary materials. As was the case with telephones, the major developers of these recycling approaches have been Western Electric and Bell Laboratories. Wehrenberg (1979) and Donovan, Pompeo, and Scalco (1977) provide details about the Western Electric/Bell Laboratories experience with recycling PVC from cable scrap. Bevis, Irvin, and Allan (1983) provide a more technical description of the various technologies available to recycle PVC and polyethylene from cable scrap. According to Donovan, Pompeo, and Scalco (1977), PVC wire contains about 63 percent copper, 25 percent PVC, with the remaining 12 percent being mostly textiles. In most of the recycling operations, the wire is first chopped into small pieces. After separation of the copper portion by taking advantage of the differences in the size and weight of the different particles, the remaining PVC and textiles are separated by air classification. Electrostatic separation may follow to remove any remaining copper particles in the ground PVC portion.

Both manufacturing and postconsumer cable scrap have been used in recycling operations. Products from the recycled thermoplastics include flower pots, irrigation pipes, auto parts, and other items that do not demand exacting properties. While there is some mention of using the recycled thermoplastics in new wiring – i.e., primary recycling – most of the waste is used in a secondary sense. It is estimated that in 1984 a total of 349 million pounds of PVC entered the postconsumer waste stream from the electrical and electronics sector. PVC waste from all manufacturers totaled an estimated 409 million pounds. Postconsumer polyethylene waste from the electrical and electronics sector is estimated at 450 million pounds in 1984. Polyethylene waste from all manufacturers is estimated at 917 million pounds for that same year. (See Appendix B for details.)

Electric Circuit Board

A more recent development in the electrical and electronics sector is the recycling of epoxy from electronic circuit boards. Rosett (1983) describes the recycling of epoxy, which is a thermoset, as a reinforcement/additive material. According to Rosett, the manufacture of circuit boards results in substantial waste tonnage from edge trim and from off-spec board. This waste material has typically been disposed by landfill. However Rosett reports that when ground, the waste can serve as an effective additive with other thermoplastic and thermosetting resins. While the composition of materials in circuit boards varies among different manufacturers, the board commonly contains between 35 and 50 percent epoxy, with the remainder composed of glass. It is estimated in Appendix B that all manufactures produced about 23 million pounds of waste epoxy in 1984. About 66 million pounds of epoxy are estimated to have entered the postconsumer waste stream from the electrical and electronics sector in 1984.

Overcoming Technological Obstacles

From a technological perspective, the recycling of plastics in the electrical and electronics sector has not required or used any revolutionary technical break-throughs. The major problem in recycling plastics as a relatively uncontaminated waste – obtaining a reliable and relatively clean waste stream – has been overcome by limiting recycling to those areas where clean waste is readily available. In all three examples the waste plastics are composed predominantly of one resin – ABS in the case of telephones, PVC or polyethylene in the case of scrap cable, and epoxy in the case of circuit board. Further, little effort is required to prevent the contamination of the waste. As discussed in the next subsection, the institutional arrangements have provided the relatively clean waste stream, rather than revolutionary developments in separation technology. The separation that has been required has been accomplished, for the most part, by using conventional technologies – i.e., air classification, magnetic separation, and sink/float methods.

In all three examples the waste plastics are being recycled in a secondary sense. Again, the technological problems posed by secondary recycling have been overcome, for the most part, by using conventional technologies. Recall from Chapter 2 that there are four basic approaches to secondary recycling: the use of conventional equipment to melt and reform thermoplastics; the use of additives to alter the chemical and physical properties of the waste resins; co-extrusion with virgin resins; and the use of waste resins as fillers with virgin resins. The first two approaches have been used in the cases of telephones and scrap cable. The recycling of epoxy from circuit boards uses the fourth approach.

Therefore, while the technical fixes that have been employed in this sector are commendable, the historical and current growth of plastics recycling in the electrical and electronics sector is not the result of major technological advances. The growth is more the result of the intelligent application of conventional technologies to selected favorable waste streams. The growth is also attributable to favorable economic and institutional conditions.

Economic and Institutional Issues

Recall from Chapter 3 that the viability of plastics recycling is largely determined by the economic and institutional incentives and barriers faced by four private-sector actors: manufacturers of plastic resins and products containing plastic parts, consumers, waste processors, and recyclers of items that contain plastic components. Further, recall that different decision makers face different incentives and barriers. In order for recycling to be accomplished outside of the municipal waste stream, not only must the required technology be available; recycling must also be economically attractive and institutional barriers must not be insurmountable. The examples of recycling in the electrical and electronics sector offer an interesting array of economic and institutional conditions that have facilitated recycling over disposal.

The decision to recycle scrap telephones has, for the most part, been due to incentives intrinsic to the service provided by the large telephone companies. Historically, most telephone service in the U.S., including the telephone equipment, has been provided by the Bell System. Until recently, the equipment used by households was not generally owned by those households, but rather was owned and serviced by the telephone company. When equipment failed there was an institutional arrangement whereby the scrap phones were diverted from the municipal waste stream. The decision to recycle was not imposed on households, because those households did not own the defective equipment.

Recall from Chapter 3 that manufacturers conceptually play two roles in the decision to recycle: one, as a producer of manufacturing waste (which is similar to the role the telephone company plays in disposing scrap telephones); and another as a designer and producer of products that contain plastic components. The telephone company plays both roles in the case of telephone recycling. (Tables 3.1 and 3.2 in Chapter 3 summarize the various incentives and barriers faced by manufacturers in both roles.)

The manufacturer's two main incentives to recycle in their role as waste processors are the avoidance of disposal costs and goodwill generated by the recycling activity. As discussed in Chapter 5, the costs of disposal are not trivial, and there is no doubt that a certain degree of goodwill has been generated by a company's recycling program. However, the most interesting observations come on the barrier side. Recall that the main barriers that face any potential recycler are technological, market, and regulatory uncertainties. In the case of telephone recycling, technological uncertainties have been minimized by using rather conventional and well established processes normally used in the processing of virgin resins. Market uncertainties are also minimal. The Bell System can forecast fairly accurately the quantities and qualities of waste that will enter the waste stream over time. Concerns about potential bias against recycled plastics are minimal because, in most cases, the recycled ABS is used in products consumed by the company. Moreover, the Bell System has had little difficult in establishing distribution channels for their products, again because the company is, for the most part, the consumer of the recycled goods. Regulatory uncertainties would also appear to be minimal at the time of the development of the recycling program. One has to wonder, however, what impact the recent breakup of the Bell System will have on the recyclability of telephones. Certainly the development of distribution channels to collect defective equipment will be more difficult. As consumers tend to purchase their own telephone equipment and as the relative prices of that equipment decrease, it is more likely that consumers will dispose of that equipment in the municipal waste stream.

In the Bell System's role as the manufacturer of a product that contains plastic components, we observe two design decisions that have assisted the recycling of telephones. First, ABS has been used consistently as the resin of choice in the manufacture of telephone bodies. This has obviously reduced the contamination of the waste being recycled. Further, according to Wehrenberg (1982), telephone housings are now marked with a code that identifies the materials used, the date molded, and the molding location. This information will assist in the future hand separation of different resins.

The economic and institutional incentives that face recyclers of thermoplastics from scrap wire and cable, and epoxy from circuit boards, are similar to those faced by the Bell System in its decision to recycle telephone bodies — i.e., the reduction of disposal costs and the generation of goodwill. However, the barriers are not so easily overcome in the cases of scrap cable and circuit boards. Potential technological uncertainties will arise from the contamination of wire insulation and sheathing with bits of copper and other materials difficult to remove in the separation step. In the case of epoxy from circuit boards, uncertainty may exist about the physical and chemical properties of composites made from mixing these ground thermosets with virgin resins. Potential consumers of the recycled products may also perceive the products to be of inferior quality. Further, to the extent the recycled goods are not consumed within the recycling organization, the distribution of the recycled goods will be more difficult.

PET BOTTLES

Background and Technological Issues

There is probably no single example of plastics recycling that has received more attention in recent years than the recycling of PET (polyethylene terephthalate) beverage bottles. The use of PET in soft drink beverage bottles was introduced in 1978 and by 1984 more than 21 percent of the total soft drink gallonage was packaged in PET bottles, according to Bennett (1985). *Chemical and Engineering News* (1981) estimates that the production of PET bottles totaled 300 million pounds in 1980. Dunphy (1985) estimates that the total usage of PET in bottles increased to 559 million pounds in 1983, of which 490 million pounds were used for soft drinks. PET bottles are also used to a much lesser extent for liquor, wine, and beer. Estimates given in Appendix B of this book indicate that the flow of all postconsumer thermopolyester waste (of which PET is a major component) from the packaging sector totaled about 595 million pounds in 1984. Projections indicate that postconsumer thermopolyesters from the packaging sector will increase to 919 million pounds by 1995.

The use of PET in bottling has increased dramatically because of several reasons. First, on a weight basis PET is drastically lighter than glass, especially when used in the new and popular one, two, and three liter containers. Second, PET is virtually unbreakable. Third, PET unlike many plastic resins resists permeation of carbon dioxide, oxygen, and water vapor. Fourth, relatively low-cost blow molding techniques have given PET an advantage in the race with glass and metal containers.

However, the recent interest in the recycling of PET bottles has not necessarily been a function of the rapid growth in the use of PET. Rather, interest has been spurred by the bottle deposit laws in at least nine states. A national bottle deposit law has been introduced and bottle laws are being considered in numerous additional states.[2]

The existing laws have provided a large and relatively clean source of PET, which, in turn, has opened numerous recycling possibilities. [Technomic Consultants (1981) have determined that the volume of recycled PET from states without bottle deposit laws is insignificant.] *Modern Plastics* (1980a) concluded that the compliance of consumers with the existing laws is generally high and approaches 95 percent in some states. Roth (1985) reports that reclamation rates average 90 percent in states with deposit laws. Dunphy (1985) estimates that about 20 percent of the 17 billion plastic soft drink bottles disposed in the United States in 1984 were recycled. About 90 percent of all of these bottles were made of PET. Bennett (1985) projects that approximately 110 million pounds of PET bottles will be recycled in 1985, which will be about three times as many as recycled in 1982.

The recycling of PET has taken many different forms, reflecting the versatility of the material. Secondary recycling methods have dominated; however, there are current examples of both tertiary and quaternary operations. As discussed in Chapters 2 and 5 of this book, one of the most noted secondary processes has been developed and demonstrated by the Goodyear Tire and Rubber Company. The technology is relatively simple and employs commercially available equipment and parts. Essentially, the process segregates the PET portion of the bottles from the aluminum caps, the paper used for labels, and the high-density polyethylene (HDPE) often used for the bases of the bottles. After grinding, the bottles are subjected to air classification and sink/float separation techniques to produce a final product that contains less than 0.2 percent impurities, according to *Modern Plastics* (1980b). While there are no regulations preventing the use of recycled PET for the manufacture of food packaging, it is generally recommended that the PET not be used for that purpose because of the remaining contaminants. Goodyear has demonstrated the process in its pilot plant in Akron, Ohio and has made the process available to all interested parties.[3]

Other technologies have been developed to separate PET from the other components of the bottles. For example, Du Pont Company has developed and used a separation process thought to be somewhat similar to the Goodyear process. (See *Modern Plastics*, 1980b.) Numerous small companies have been formed to recycle PET in a secondary sense, such as the St. Jude Polymer Corporation located in Frackville, Pennsylvania, which according to *The Plastics Bottle Reporter* (1985a) has the capability to process as much as 18 million pounds of PET per year.

Numerous secondary products have been produced from recycled PET. *Plastics World* (1984) reports that perhaps 75 percent of all recycled PET is currently being used for fiberfill in pillows, coats, sleeping bags, and so forth. Yacona (undated) reports that 9 two-liter bottles produce about one pound of fiberfill. Five bottles are sufficient for one man's ski jacket. A sleeping bag can be filled with about 36 two-liter bottles.

In the second largest recycling use, PET is chemically altered by a tertiary process to produce unsaturated polyester from which an assortment of products can be produced. The polyester obtained from the recycled PET can be used in fiberglass to manufacture products such as bathtubs and shower stalls. PET can also be used to produce industrial strapping and even sailboats. Tertiary processes include those by Eastman Chemical Products (see *Modern Plastics*, 1980b) and a process developed by Michigan Tech University (see Barna, Johnsrud, and Lamparter, 1980). Most tertiary processes will require that the bottles be subjected to a preliminary separation step, such as those discussed in the previous paragraph.

Recycled PET has also been used in a quaternary sense to retrieve its heat energy of approximately 9,500 Btus per pound. (The bottles, including the HDPE base, may produce between 10,000 and 12,000 Btus per pound according to *The Plastic Bottle Reporter*, 1983.) An example of this technology is the rotary kiln incinerator with heat recovery developed by Industronics Incorporated. [See Industronics, Inc. (1982) or *The Plastics Bottle Reporter* (1983) for details.] The process reportedly can burn whole bottles without shredding or the removal of the aluminum caps, HDPE base cups, and labeling. Further, air pollution is said to be within government limits. *Chemical Engineering* (1982) reports that recycled PET has also been used as a supplemental boiler fuel in an Oregon paper facility.

Therefore, a host of technological approaches have been used to recycle PET. However, the technologies currently being used are, for the most part, adaptations of conventionally used equipment. The major technological problem, that of obtaining relatively uncontaminated waste, has been solved by the economic and institutional environment created by current bottle deposit laws. Separation of the PET from the other components of the bottle has been rather straightforward and has employed well known air classification and sink/float methods. Further, the most successful use of recycled PET, that of fiberfill, has not required the development of new and sophisticated technologies. The PET bottle is a good example of a waste stream that meshes nicely with technologies that have been developed for other purposes.

Economic and Institutional Issues

The recycling of PET bottles is a case in which public sector incentives have been the key to success. Given that an insignificant quantity of PET is being recycled from states without bottle deposit laws, we can conclude that the recycling of PET would not be extensive in the absence of those laws. The 5–10 cent deposits in the states with such laws, in combination with individual concerns about the disposal of PET bottles, have been sufficient incentives to overcome the consumer barriers to recycling, i.e., the inconvenience of source separation and storage and the cost of transportation to a deposit refund center. Compliance rates have exceeded 90 percent in many states with bottle laws.

Once this major obstacle of diverting the PET bottles from the municipal waste stream has been overcome, recycling in some form becomes the expected. Two reasons dominate. First, economic and institutional incentives encourage the collectors of the waste to recycle. Nontrivial disposal costs must be paid if some form of recycling cannot be found. And in the case of PET bottles, many technological fixes are available with minimal technological uncertainties. In addition, market uncertainties are minimal in that the bottle laws assure supplies of waste materials. Numerous market opportunities exist for clean PET. And any consumer bias against the recycled product is often masked because the PET is either converted to an unsaturated polyester from which new products are made or is used as fiberfill in products where the recycled material is not visible to the consumer.

Second, the manufacturers of PET bottles have an incentive to assist in the development of recycling technologies. Given that plastic bottles are being diverted from the municipal waste stream, it is in the manufacturer's interest to make PET as valuable as competing glass and metal materials. Goodyear's activity in this area is representative of the industry's resolve to increase the value of waste PET bottles by offering a proven recycling technology to interested parties. Another example is the work on PET recycling by the industry-supported Society of the Plastics Industry and the recently formed Plastics Recycling Institute.

PLASTICS IN AUTOMOBILE SHREDDER RESIDUE

Background and Technological Issues

Reasons for Interest

A good example of a waste stream that does not currently recycle plastics but has an excellent chance of some form of recycling in future years is residue from automobile shredder operations. A significant amount of work is currently being done by both the public and private sectors to develop recycling processes that will be suitable for this waste stream.

Interest in shredder residue has been driven by three key forces. First, of the approximate 25 percent of all plastic wastes that can realistically be diverted from the solid waste stream in 1990 (projected to be 8.4 billion pounds), about 27 percent of the total (or about 2.2 billion pounds) will come from the transportation sector.[4]

Second, there has been a rapid shift in the technology used to process scrap automobiles. Prior to the mid 1960s the vast majority of scrap automobiles were processed by a combination of hand dismantling and sorting. The remaining predominantly steel hulks — which contained various contaminants, including plastics — were then "baled" into compact bundles to facilitate handling and transportation. Using the baler technology, the plastic portion of the scrap automobile is not separated from the metallic portion and thus is lost from the recycling stream. However, the introduction of the shredder technology in 1963 allowed the scrap processor first to shred autos into fist-sized pieces that could be separated into ferrous, nonferrous, and nonmetallic portions. This additional separation resulted in the production of much superior scrap steel, allowing the automobile recycler to enter the No.1 steel scrap market, rather than sell their product as lower-priced No.2 steel scrap.[5] However the shredders also resulted in an accumulation of an unwanted nonmetallic waste that has typically been disposed by landfill at the auto recycler's expense. It has been estimated that the nonmetallic portion of shredder residue consists of between 14 and 36 percent plastics. [See Metal Scrap Research and Education Foundation (1983). More detailed information on the composition of shredder residue is given below.] The nonferrous portion, which is typically processed further to retrieve zinc, aluminum, and other valuable metals, has been estimated to contain about 21 percent plastics. (See Valdez, 1976.) The most recent estimate (for 1981) of total shredder capacity in the United States was placed at 13,400,500 tons per year.[6] In 1982 about 75 percent of all recycled autos were processed using shredders.[7]

The third driving force is the historical evidence that manufacturers of automobiles have increased the use of plastics in automobiles and the widely held belief that plastics will compose an even larger percentage of the auto's weight in future years. The use of plastics in automobiles has increased from less than 1 percent of the total weight of the typical domestic auto in 1960 to about 6.4 percent in 1982 — or about 200 pounds of the 3,114 pound weight of the average U.S. automobile (Holcomb and Koshy, 1984). The *American Metal Market* (1984b) reports that the Ford Motor Company expects their 1992 cars to contain 270 pounds of plastics, which is a 14 percent increase over their 1984 models. U.S. automobile manufactures consumed more than 1 billion pounds of plastics in 1983 and are projected to consume 2.2 billion pounds by 1995 according to the *American Metal Market* (1984a). During this time period plastics in exterior components are expected to increase from 21 to about 31 percent of the total plastics used in autos. Mechanical and structural plastic components will remain at about a constant percentage level. Plastics in interior uses are projected to decrease in percentage terms.[8] Projections presented in Chapter 4 and Appendix B of this book indicate the total use of plastics in the domestic transportation sector will be about 2.2 billion pounds in 1990 and 1995.

Many experts predict that the content of plastics in future automobiles will increase due to the automakers' continued drive to reduce auto weight and thus improve fuel efficiency.[9] Bever (1980) reports projections of the composition of automobiles produced in 1990 assuming extreme use of three alternative light-weight materials: high-strength steels, aluminum, and plastics. The projected consumption of plastics ranges from a low of about 10 percent under the case of the extreme application of aluminum to about 25 percent under the extreme application of plastics. Therefore, even under the case of the extreme application of aluminum, Bever projects that the consumption of plastics in automobile manufacture will increase sharply above the current level.

Potential Recycling Processes

A significant amount of work is currently being done to develop processes that could utilize the plastics in shredder residue. The Oak Ridge National Laboratory, as part of the U.S. Department of Energy's Energy Conversion and Utilization Technologies Program and in association with various universities and the Plastics Institute of America, is currently doing work to develop processes that would use cleaned and ground shredder residue to produce composite materials. The processes would include mixing plastic binders, such as melamine and phenolic resins, with the residue to increase the strength and impact resistance of the composite material. Numerous products could be produced, including floor coverings, bulky items such as drainage gutters and flower pots, and a weather resistant particle board. It is believed that the particle board would have strength characteristics similar to particle board made from wood.[10]

Work has also been done by the Ford Motor Company to develop a method for the hydrolysis of the polyurethane foam from shredder residue. The resulting material could be used for the synthesis of new industrial chemicals. For additional information on the work by Ford and other works in this area see Harwood (1977a and 1977b) and Bever (1980). Yet another possibility for shredder residue is in concrete as a replacement for sand and gravel. The resulting concrete could potentially be more resistant to variations in temperature and moisture conditions.[11]

Some Keys to Technical Success

The technical and economic feasibility of using the technologies discussed above and many of the secondary and tertiary technologies discussed in Chapter 2 to process residue from shredder operations depends crucially on the composition of the residue and the ability to decontaminate the plastics in that waste stream to an acceptable level. The composition of the residue, in turn, depends on the quantities of specific resins and other nonmetallic materials historically used in automobiles and on the ways those nonmetallics are separated following the shredding operation.

Plastic Wastes from the Transportation Sector.   The current life of the average domestic automobile is 10.9 years. Given historical information on the use of specific resins in the transportation sector, Chapter 4 and Appendix B present information on the projected flows of specific resins from the transportation sector. Reviewing those results briefly, we find that of the estimated 1.9 billion pounds of postconsumer plastics from the transportation sector in 1984, 29.9 percent were thermosets, 18.2 percent were polyurethane foams, and the remaining 51.7 percent were thermoplastics. The vast majority of thermosets was polyester at 24.2 percent of the total. Polypropylene was the largest thermoplastic contributor at 17.0 percent, followed by ABS and SAN, and PVC at 12.7 and 12.1 percent of the total, respectively. Projections for 1995 indicate that total postconsumer waste from the transportation sector will increase to 2.3 billion pounds. Thermosets will decrease in percentage terms to about 23.2 percent of the total, with polyester remaining largest at 14.4 percent and phenolics second at 7.1 percent. Polyurethane foams will compose about 13.9 percent of the total. Thermoplastics are projected to increase to 62.9 percent of the total, with polypropylene contributing 13.3 percent, ABS and SAN 12.1 percent, and PVC 6.8 percent.

The trends toward the use of more thermoplastics in automobiles is an incentive to recycle. However, there is evidence to suggest that plastics will be used more in composite materials to increase the overall strength of the components. Reinforced plastics are now being used in some domestic models for body panels, and recent reports project that the use of reinforced plastics and composites will increase rapidly in future years. As is discussed in Bever (1980), the glass or graphite fibers used in these composite materials may cause these reinforced plastics to be unacceptable for certain types of recycling.

Plastics in Shredder Residue.   Plastics are only one of several materials found in shredder residue, which is a serious barrier to recycling. As stated earlier, residue from shredders is separated into nonferrous and nonmetallic portions, and both portions contain relatively high percentages of plastics. The non-ferrous portion contains significant quantities of zinc, aluminum, and copper.

To recover these metals, this portion is typically shipped from shredders to a central location where additional processing can take place. The plastics and other materials remaining after this additional processing step have typically been landfilled. The nonmetallic portion has typically been landfilled by the shredder operator.

At least three estimates of the contents of shredder residue have been made. Mahoney, Braslaw, and Harwood (1979) found that the nonmetallic portion from a 1972 model contained virtually all of the polyurethane foam, 25 percent of the non-foam plastics, 16 percent of the rubber, and 6 percent of the nonferrous metals. The nonferrous portion was found to contain 80 percent of the nonferrous metals, 45 percent of the non-foam plastics, 45 percent of the rubber, and less than 1 percent of iron and steel. The remaining materials were in a nonrecoverable form in the nonferrous and nonmetallic portions — i.e., they were too small for recovery — and were found as contaminants in the ferrous portion.

The Metal Scrap Research and Education Foundation (1983) gives a detailed breakdown of the contents of four samples of nonmetallic scrap after the removal of dirt, stones, and glass. "Plastics" composed an average of 19.3 percent of this residue portion by weight, ranging from a high of 36.4 percent to a low of 12.6 percent.[12] Fibrous and cloth materials composed an average of 42.1 percent, paper 6.5 percent, and foam 2.3 percent. Approximately 86.4 percent of the residue was composed of organic materials, with the remainder consisting of glass, metals, and electrical wire.

Valdez (1976) presents a relatively detailed breakdown of the composition of both nonferrous and nonmetallic residue prior to the removal of dirt, stones, and other miscellaneous materials. The sample residue came from the shredding of five GMC cars and one Volkswagen produced during the mid 1970s. The results, summarized in Table 6.1, show that plastics (exclusive of foam and fibers) composed about 15.9 and 13.7 percent of nonferrous and nonmetallic residues, respectively. Inclusive of foam and fibers, the nonferrous and nonmetallic residues contained 21.0 and 41.8 percent plastics, respectively (under the assumption that all fibers are made from plastic resins). It is interesting to note that a large percentage of both residue portions is composed of dirt and other nondistinguishable materials — 23.6 percent in the case of nonferrous residue and 34.6 percent in the case of nonmetallic residue.

The results of these studies indicate that a significant barrier to the successful recycling of plastics from shredder residue will be the additional separation that will typically be necessary to obtain a stream of relatively uncontaminated plastics. Because of the contaminants in shredder residue, tertiary and quaternary recycling processes may have a relative advantage over proposed secondary processes that require less contaminated waste streams. Work has, however, been done to separate plastics from the dirt and other materials in the residue portion with some degree of success. Most processes

TABLE 6.1. The Composition of Nonferrous and Nonmetallic Shredder Residue by Weight (in percentages)

Material	Nonferrous Residue	Nonmetallic Residue
Zinc	12.3	0.2
Magnesium	3.5	0.5
Copper	4.6	0.7
Steel	2.1	0.5
Aluminum	12.7	0.8
Lead	0.4	0.0
Iron	2.3	9.4
Plastics	15.9	13.7
Foam	0.0	11.7
Rubber	10.8	2.1
Fiber	5.1	16.4
Paper	1.3	4.6
Wood	0.4	0.0
Glass	5.0	4.7
Dirt	4.6	7.9
Other	19.0	26.7
Total*	100	100

*May not sum to 100 because of rounding.
Source: Valdez (1976); compiled from information given in Tables 3 and 4, pp. 389 and 390.

include washing followed by density fractionation by air and water mediums. For more information on these separation approaches see, for example, Deanin and Nadkarni (1984), Dreissen and Basten (1976), Valdez (1976), and Valdez, Dean, Bilbrey, and Mahoney (1975).

Economic and Institutional Issues

The potential recycling of automobile shredder residue is a case in which many economic and institutional incentives and barriers come into play. Consumers play a role in that they have a monetary incentive to divert scrap autos to recyclers so that the autos' steel content can be retrieved. Automobile manufacturers have impacted the decision to recycle by generally increasing the percentage of plastics in the typical automobile. While this trend on the part of manufacturers has been an incentive to recycle, the manufacturer's current movement toward the use of plastics in composite materials may be a serious barrier. Another barrier introduced by manufacturers is the use of many different resins in different automotive parts, which may result in mixtures having poor physical and chemical characteristics. The incentives for automobile manufacturers to adopt these changes have mainly come from demands by consumers and government to increase fuel economy. Further, new and cheaper processes to manufacture plastic components have made plastics competitive with more traditionally used materials.

In addition to mandating fuel efficiency standards, the public sector has played a key role in promoting recycling by providing support for the development of new technologies to recycle shredder residue. The support by the public sector can be defended on the grounds that plastics from the transportation sector compose a large percentage of all postconsumer plastics that can realistically be diverted from the municipal waste stream. If the public sector is going to be active in promoting the recycling of plastics outside of the municipal waste stream, the recycling of shredder residue is a logical place to begin because of that waste stream's institutional and technical characteristics.

The most important player in the decision to recycle is, however, the shredder operator. From the perspective of the shredder operator and the operator of the nonferrous recovery plant who must dispose of their residue by landfill, the primary incentive to adopt a recycling process is clearly to reduce or eliminate the cost of disposal. If recycling is expected to cost less than disposal, an economic incentive will exist.

Table 5.7 in Chapter 5 presents the expected direct costs and revenues associated with one proposed recycling technology. Recall that the profitability of that operation is expected to be quite sensitive to operational size. Table 6.2 presents information on the cost of landfill for auto shredders in different population regions of the United States. Information on the cost of landfill by population size has been combined with information on the sizes and locations of domestic shredder operations to suggest the financial incentives of shredder operators to adopt recycling.

It is estimated that shredders in 1981 were paying between $5.61 and $15.33 per ton to dispose of shredder residue. Note that disposal costs generally increase with population size. Table 6.2 also indicates that about 59.3 percent of all domestic shredders are located in areas where the cost of disposal is in excess of $11 per ton in 1981 dollars. Therefore, given these findings, a recycling process that can utilize shredder residue for a cost of less than $11 per ton (in 1981 dollars) will be economically attractive to about 60 percent of the shredder market. In order for the process to be attractive to the total market, the cost of recycling will have to be in the $5 to $7 range. (See Table C.1 in Appendix C for a complete listing of shredder facilities arranged by population size and capacity.)

There are, unfortunately, several economic and institutional barriers that impact on the potential recycler. First, the technologies that are currently

TABLE 6.2. U.S. Shredder Capacity by Population Size and Cost of Landfill

Size of Population Region (thousands)	1981 Shredder Capacity (thousands tons/year)	Percent of Total	Landfill Cost (1981 $/ton)
>500	2,946.5	22.0	13.54
250–500	2,253.0	16.8	15.33
100–250	2,744.0	20.5	11.80
50–100	2,262.0	16.9	7.59
25–50	1,242.0	9.5	5.61
10–25	1,387.0	10.4	8.32
2.5–10	566.0	4.2	6.98
Total	13,400.5	100*	

*Does not sum to 100 because of rounding.

Sources: Shredder locations and operational sizes: *Scrap Age* (1980); population data: Andriot (1983); landfill cost by population size: U.S. Environmental Protection Agency (1984)

available have not been demonstrated on a large scale. Second, market uncertainties arise on several fronts. The production of shredded steel scrap — and therefore shredder residue — has historically been volatile, reflecting large variations in prices for scrap steel. [See Curlee (1985b) for details.] Recyclers must therefore be concerned that residue will not be available in sufficient quantities over time to justify a processing plant. In the cases where shredders have small capacities, residue from several sites may have to be pooled to justify a recycling operation. Transportation costs may prove a major barrier if residue must be transported a long distance. Finally, recyclers may find their products are stigmatized because they are made from recycled materials.

CONCLUSIONS

This chapter has reviewed three market sectors where plastics recycling is either currently taking place or has a good chance of being implemented in future years — i.e., the electrical and electronics sector, PET beverage bottles, and residue from automobile shredder operations. The discussion of these sectors is important because technological assessments of recycling have often focused on these product areas. However, at a more fundamental level, the assessment of these areas indicates the types of technical, economic, and institutional conditions that promote recycling outside of the municipal waste stream.

From a technical perspective the overriding conclusion is that recycling has been most successful where conventional technologies or slight variations of those technologies can be used. For the most part, the recycling of telephones and plastics from electrical cable have employed commonly used equipment. Further, only limited separation has been required because of economic and institutional incentives that have provided relatively clean scrap. Technical problems resulting from waste contamination and mixtures of various plastic resins pose a major obstacle to the recycling of auto shredder residue.

The major conclusion of this chapter, however, is that successful recycling operations have been influenced more by favorable economic and institutional conditions than by technological fixes to the problem. The major problem in recycling plastics outside of the municipal waste stream—i.e., obtaining a relatively clean waste—has been overcome by economic and institutional incentives. In the case of telephones, the historical arrangement of lease/service has diverted scrap equipment from the municipal waste stream. In the cases of electrical cable and shredder residue, diversion has resulted from a desire to recycle metals. And in the case of PET bottles, diversion has resulted from direct intervention by the public sector. Further, in each of our examples, incentives other than the desire to recycle plastics have contributed to the flow of a relatively clean plastic waste stream. Plastics in telephones are partially separated from other materials in an attempt to repair the equipment. Plastics are rejected as a by-product in the recycling of electrical cable and automobiles. And in the case of PET bottles, the public sector's desire to divert all beverage containers from the general waste stream has resulted in a relatively clean source of PET.

If the currently successful recycling operations indicate the types of recycling we can expect in future years, those operations will not develop around revolutionary technological advances. Rather, they will result from the resourceful application of existing technologies to relatively clean waste streams, made so by an assortment of economic and institutional incentives.

NOTES

1. For more information on the recycling of ABS from scrap telephones see, for example, Wehrenberg (1979 and 1982), Criner (1977), and Hancock and Hubbauer (1975).

2. As of 1985, states with bottle deposit laws included Delaware, Massachusetts, New York, Oregon, Connecticut, Iowa, Maine, Michigan, and Vermont. The first bottle deposit law was enacted in Oregon in 1971. A national deposit law was introduced by Senator Mark Hatfield of Oregon in 1983 but did not achieve passage.

3. For more information on the Goodyear process see, for example, Goodyear (undated), *Chemical and Engineering News* (1981), *Journal of Commerce* (1980), *Modern Plastics* (1980a), and *Waste Age* (1979).

4. The content of this section is derived in part from previous work by this author; see Curlee (1985a, b, c).

5. The majority of all scrapped automobiles — between 80 and 97.5 percent, depending on the source cited — have been recycled predominantly for their steel content. Kaiser, Wasson, and Daniels (1977) estimate that between 82.2 and 97.5 percent of all vehicles deregistered between 1958 and 1975 have been processed into scrap. The balance of the scrapped vehicles have been landfilled, otherwise disposed, or remain in the stockpile of potential auto scrap. Harwood (1977a) bounds the estimate of auto scrapping at between 80 and 85 percent, while Ewert (1976) estimates that between 85 and 88 percent of retired vehicles enter some part of the scrap cycle. Robert E. Nathan Associates (1982) estimate that during 1972–81 an average of 17.3 million tons per year of recoverable automotive obsolete scrap was generated.

6. This estimate is based on data from *Scrap Age* (1980). The 1981 estimate of shredder capacity is based on estimates of planned additional capacity as of 1980.

7. Curlee (1985b) contains a detailed analysis of the historical adoption of automobile shredders. Also discussed in that report is an econometric approach to measure the degree to which a new technology has replaced an existing technology in a given market. Using that approach, the degree to which shredders have penetrated the automobile recycling market is assessed. It is concluded that while the market for scrap steel has not been saturated by the shredder technology, the movement toward shredders and away from balers has slowed significantly.

8. For additional information on the use of plastics in automobiles see, for example, *Machine Design* (1984), Berry (1983), and *Chemical Week* (1982a and 1982c).

9. According to Holcomb and Koshy (1984) the typical 1975 model U.S.-built automobile contained 59 percent steel, 15 percent iron, 4 percent plastics, 2 percent aluminum, 2 percent glass, and 18 percent other materials. In 1983 those percentages had changed to 54 percent steel, 15 percent iron, 6 percent plastics, 4 percent aluminum, 3 percent glass, and 18 percent other materials.

10. See Chapter 5 for estimates of the expected costs and revenues associated with a recycling process to produce composite material from shredder residue.

11. For more information on the work to develop a composite material from automobile shredder residue see *Machine Design* (1984) and Plastics Institute of America (draft).

12. This definition of plastics does not include fibrous plastics, polyurethane foams, or elastomers.

7
Conclusions

INTRODUCTION

In the course of this book a broad spectrum of technical, economic, and institutional questions have been posed and to some extent answered. It has been argued that the technical problems presented by plastics recycling are often overshadowed by economic and institutional issues. In addition, the relevant economic and institutional issues must be defined within the context of what is technically feasible. A failure to mesh the technical, economic, and institutional issues will result in a partial analysis at best and may result in misleading conclusions about the future recyclability of plastic wastes.

The purpose of this final chapter is not simply to summarize the numerous conclusions drawn in preceding chapters, but rather to apply those conclusions in an attempt to answer several general questions, the answers to which will largely determine if future plastic wastes will be recycled or disposed of. Although the answers are incomplete given our current state of knowledge, this discussion will hopefully help to define the relevant questions to be addressed in future assessments of the public and private sectors' decisions about plastics recycling.

IS THE RECYCLING OF PLASTIC WASTES
REALISTIC FROM A TECHNICAL PERSPECTIVE?

The answer to this question hinges on how one defines recycling. As discussed predominantly in Chapter 2, plastics recycling can take several forms: primary recycling, secondary recycling, and tertiary or quaternary recycling as part of the municipal waste stream or as a segregated waste. If one takes this broad perspective of recycling, then all plastic wastes are in a technical sense currently recyclable in one form or another. If one takes the more narrow perspective that recycling means only primary or secondary methods, then most plastic wastes are not currently recyclable nor are they expected to be recyclable in the near future.

The more relevant question concerns how technical constraints limit the application of currently available or developmental recycling technologies to specific waste streams. Two main technical constraints arise: a) variations in the physical and chemical properties of different resins; and b) current difficulties in separating plastics from other similar materials and in separating specific resins. Unfortunately, from the perspective of recycling, the dissimilar physical and chemical properties exhibited by different resins pose significant problems in the primary, secondary, and sometimes tertiary recycling of plastics. We can broadly categorize resins into thermoplastics and thermosets and conclude that a mixture of different thermoplastics is more recyclable than a mixture of thermosets and thermoplastics. However, the different properties exhibited by different thermoplastics prohibit the primary recycling of a mixture of those resins and limit the applications of products produced by a secondary process. Further, while tertiary recycling is more applicable to a mixture of resins, some currently available tertiary processes are designed to process only one resin.

The technical problems posed by mixtures of different resins could, of course, be overcome if separation is a reasonable goal. However, as has been discussed in some detail in Chapter 2, the general conclusion from the technical literature on separation is that the separation of plastics from similar materials, such as paper, is difficult, and the separation of different resins is not currently a viable option. Technologies do exist that perform these functions and have been successfully demonstrated. However, their technical complexity and relatively high cost have led the technical experts to conclude that the existing separation processes will, for the most part, be limited to very selective uses in the coming decade. It is not anticipated that plastics will be separated from other similar materials in the municipal waste stream, much less segregated into specific resin types. Similarly, for plastic wastes collected outside of the municipal waste stream, the major obstacle to recycling will be the ease with which plastics can be separated from contaminating materials.

Therefore, from a technical perspective the following general conclusions can be drawn. First, the plastic wastes that are currently being disposed of are in virtually all cases not recyclable in a primary sense because the wastes are contaminated with other materials or because the physical and chemical properties of the waste resins pose insurmountable barriers.

Second, secondary recycling will be limited to plastic wastes that can easily be diverted from the municipal waste stream either as a single resin or as a mixture of resins. It is generally concluded that plastics that enter the municipal waste stream will not be recyclable in a secondary sense in the near future because of the arduous task of materials separation. However, for those waste streams that can be collected separately from other wastes, several secondary processes are currently available, and other processes are in the developmental stage. These technologies, examples of which are applicable to both thermoplastics and thermosets, are often variations of equipment commonly used in the processing of virgin resins. In most cases, thermoplastics are melted and reformed into new products with properties less demanding than those of the original products. Because of their interlinking molecular bonds that prevent melting, thermosets are ground and used as fillers with virgin resins.

Third, for those plastic wastes that can realistically be diverted from the municipal waste stream, tertiary and quaternary recycling processes are currently available and are being used commercially. These processes, which require relatively clean waste and generally produce a high-value product, offer an attractive alternative to the secondary processes mentioned above.

Finally, for those waste plastics that cannot be diverted from the municipal waste stream, recycling possibilities exist in the forms of tertiary and quaternary recycling with other waste materials. These processes have been tested thoroughly and in the case of quaternary recycling are used extensively throughout the United States. Pyrolysis processes, although technically feasible, have not received the same level of attention.

Plastics are particularly suited to quaternary recycling with other municipal wastes because of their high heat values — roughly equivalent to coal on a weight basis. Given that plastics compose only about six percent of the typical municipal waste stream, the technology currently exists to recycle all plastic wastes with other combustible waste materials in a quaternary sense with minimal technical problems. As the percentage of plastics in the waste stream increases, the burning of plastics with other waste materials becomes more difficult in conventional incineration equipment because of the increased demands for oxygen, the excessive heat produced, and clogging problems. Innovative technologies are, however, currently available to burn segregated plastics, such as PET bottles, and are reported to have few technical problems.

WHAT ARE THE ENVIRONMENTAL
IMPLICATIONS OF PLASTICS RECYCLING?

There is strong evidence to suggest that the general public perceives plastic waste to be harmful to the environment — by far the most harmful of all commonly used materials. The public believes that the incineration of plastics results in toxic fumes that escape to the atmosphere. Further, there is a general feeling that the nonbiodegradable characteristics of most resins make plastics unacceptable in landfills.

The technical experts do not, however, reach such a strong consensus. In the case of incineration with and without heat recovery, the experts acknowledge that some resins, such as PVC, can produce hydrogen chloride, which can in turn react with water to produce hydrochloric acid. They also acknowledge that the residue from plastics incineration can contain elements such as cadmium and lead, which are used as stabilizers in some resins. However, the experts differ in their assessments of the environmental impacts of incinerating plastics.

At one extreme position the experts argue that currently available incineration equipment can effectively counter any environmental threats posed by plastics incineration. Those in favor of incineration also argue that the high Btu values of most resins help to elevate the temperature of municipal waste incinerators, which in turn helps to incinerate waste materials that are inherently difficult to burn or contain a high moisture content. In this sense, plastics in municipal waste incineration help, rather than hurt, environmental quality. Finally, the pro position for incineration often claims that even if some environmental degradation occurs, that negative impact is minuscule compared with other environmental problems we face from other sources, such as the burning of high-sulfur coal.

The other extreme position argues that current technology cannot sufficiently counter the environmental threats posed by plastics incineration as a part of, or segregated from, the municipal waste stream. This position also argues that while equipment is often available to reduce environmental damage, the equipment is frequently used improperly or not at all.

The environmental impacts of landfilling plastic wastes are equally controversial. On the one hand, some argue that the nonbiodegradability of most resins means that plastics in landfills will remain essentially unchanged for many years. In addition, because plastics are difficult to compact, the landfill may develop spongy areas, which may detract from the usefulness of the landfill site once completed. The same characteristic may expose other degradable and potentially harmful materials to water and thus facilitate the spread of toxic substances outside of the landfill area.

On the other hand, some experts see the same characteristics as being helpful to the landfill. Because most plastics are nonbiodegradable, they do not produce harmful liquids or gases. Further, the decomposition of plastics — which does occur over a long period of time — produces substances that are usually inert. This position also argues that plastics help to provide structural support to the landfill if distributed properly. As other degradable materials decay, plastics provide structural support that facilitates the use of the site after the landfill is closed.

Unfortunately, the debate about the environmental impacts of plastics disposal is one of those issues for which most any position can be defended or attacked. In a case such as this where the level of technical uncertainty is high, the issue is inevitably reduced to the political arena where a public decision is based more on political clout than on scientific evidence.

While the debate about the environmental impacts of plastics disposal continues, an equally important and possibly more tractable question is whether recycling could eliminate or significantly reduce the alleged environmental damage that results from disposal. Recall that technical constraints allow for the secondary, tertiary, or quaternary recycling of plastics that have not entered the municipal waste stream, and the tertiary or quaternary recycling of plastic wastes as part of the municipal waste stream. The question then becomes, do these forms of recycling pose fewer environmental problems than disposal? Further, is any one form of recycling preferable to another form?

Consider each of the recycling possibilities. Recall from Chapter 2 that the secondary reuse of plastic wastes usually involves the heating and remolding of waste thermoplastics into new and usually bulky products. Minimal air and water pollution results. The problem with secondary recycling from the environmental perspective is, however, that waste resins cannot be repeatedly remelted and remolded. The quality of the resins is damaged to some extent each time they are melted and reformed, implying that a waste resin cannot be repeatedly recycled in a secondary sense. Eventually the waste must enter a tertiary or quaternary recycling process or be disposed of by incineration or landfill. Secondary recycling thus delays but does not eliminate the need for disposal or for recycling by tertiary or quaternary means. The environmental benefits of secondary recycling must therefore be measured in terms of deferring, rather than eliminating, the environmental consequences of disposal or other recycling processes. Obviously, the higher the assumed discount rate (or the assumed tradeoff between costs incurred today versus costs incurred at some future time), the more valuable the environmental benefits of secondary recycling become. Given that secondary products do not usually displace the consumption of virgin resins, but rather substitute for other materials such as wood and metal, secondary recycling will not in the long term reduce the quantity of plastics produced or discarded.

Once we eliminate secondary recycling as a long-term solution to the potential environmental problems posed by plastics disposal, we must consider the environmental implications of tertiary and quaternary recycling outside of, and as a part of, the municipal waste stream. Unfortunately, once again we enter an area where there is significant technical disagreement. Incineration with heat recovery will pose problems similar to incineration without heat recovery. And while it is generally agreed that tertiary recycling is less polluting than incineration, the environmental impacts of processes such as pyrolysis are not nil. Further, since tertiary recycling produces basic chemicals and fuels, one must consider the environmental implications of using these products as part of the environmental impacts of tertiary recycling.

Three general conclusions can be drawn. First, there is no general consensus among the technical experts about the environmental implications of plastics disposal. Second, secondary recycling delays but does not eliminate the need for, and the environmental consequences associated with, disposal, or tertiary or quaternary recycling. Third, given current information, it is difficult to compare the environmental impacts of tertiary and quaternary recycling with those of landfill and incineration without heat recovery.

HOW MUCH PLASTIC WASTE WILL
BE PRODUCED IN THE COMING DECADE?

Chapter 4 has presented in some detail projections of the quantities of future manufacturing and postconsumer plastic wastes through the year 1995. Those projections indicate that total plastic wastes will grow rapidly — increasing by about 46 percent above the 1984 level. Postconsumer and manufacturing wastes are estimated to have been 32.5 billion pounds in 1984 and are projected to total 47.4 billion pounds in 1995. The vast majority of all plastic wastes — 91–92 percent of the total — will come from the scrapping of consumer products. Postconsumer wastes in 1995 are projected to be composed of 87.8 percent thermoplastics, 7.6 percent thermosets, and 4.6 percent polyurethane foams. Thermoplastics, thermosets, and polyurethane foams are projected to account for 82.2 , 13.3, and 4.5 percent of all manufacturing nuisance plastics in 1995, respectively.

The more important consideration in an assessment of the recyclability of plastic wastes is not, however, the total quantity of waste to be produced, but rather the form in which the waste will occur. As discussed above, wastes that cannot realistically be diverted from the municipal waste stream will not be recyclable in a secondary sense and will not be applicable to tertiary and quaternary methods that require relatively clean waste.

The product categorizations used in Chapter 4 allow some rough calculations of the quantities of waste applicable to the different recycling streams. For example, some product categories, such as consumer goods and packaging will (with the exception of returnable bottles) be very difficult to divert from the general waste stream. Other product categories, such as electrical equipment and automobiles, are normally diverted from the general waste stream to retrieve their valuable metal contents. The remaining residue often contains a large percentage of plastics and could be collected independently of other solid wastes. Further, manufacturing wastes that are produced in large quantities at one location and are often limited to a small number of known resins, could offer a relatively clean source of plastics for recycling outside of the municipal waste stream.

It is argued in Chapter 4 that if all manufacturing nuisance plastics and all postconsumer plastic wastes from plastic beverage bottles, the electrical and electronics sector, the transportation sector, and the industrial machinery sector were successfully diverted from the municipal solid waste stream, those wastes would equal about 26 percent of all the plastic wastes projected to be produced in 1990. Applying the same conditions to the 1995 waste stream, about 23 percent could be diverted from the municipal waste stream. And while some waste might be diverted from other product categories, such as the construction sector where plastic wastes are expected to grow rapidly during the coming decade, it is not realistic to expect that more than about one quarter of all plastic wastes can or will be recyclable outside of the municipal waste stream.

WILL THE PRIVATE SECTOR ADOPT PLASTICS RECYCLING IN THE ABSENCE OF GOVERNMENT ASSISTANCE?

The difficulty in answering this question begins with defining a situation in which "government assistance" is not present. A major contributor to the private sector's current interest in plastics recycling is the body of environmental legislation and regulations that has elevated the cost of disposal above the now socially unacceptable practice of open dumping. The recently enacted bottle deposit laws in many states are another major contributor. Alternatively, government regulations may have hindered the movement toward recycling by, for example, limiting the siting of recycling facilities.

Let us therefore pose the question of whether the private sector will adopt recycling within the context of the current state of government involvement.

As has been discussed throughout this book, there are many examples of currently active plastics recycling operations, e.g., the secondary recycling of PET bottles, telephone equipment, and electrical cable; the tertiary recycling of PET and several industrial waste resins; and the quaternary recycling of PET. Moreover, it is likely that there will be additional private-sector movement toward the types of recycling operations that are now apparently viable alternatives to disposal.

Unfortunately, the issue of plastics recycling is far too complicated and heterogeneous to allow any general statement about the degree to which recycling will be adopted by the private sector in the coming decade. As discussed in the previous section, the inability to divert more than about 25 percent of all plastic wastes from the municipal waste stream during the 1990-95 time frame may place an upper bound on the extent to which the private sector may be involved. However, what may be an equally important and more tractable question, especially from the public's perspective, is what technical, economic, and institutional parameters will influence the private sector's decision.

A somewhat simplistic but often ignored criterion for recycling is that the benefits to all parties affected by recycling must exceed the costs imposed on those parties. And in those cases where the total benefits of recycling exceed the costs, but the costs exceed the benefits to some segment or segments of the private sector, a mechanism must exist for the redistribution of the cost and benefits such that all affected parties have a net incentive to recycle. The evaluation of the private sector's desire to recycle cannot simply be reduced to a comparison of the estimated net accounting cost of a recycling process versus that of disposal. As is discussed in detail in Chapter 3, the specific costs and benefits to different segments of the private sector, which reflect their perceived incentives and barriers to recycle, will vary widely and assume various monetary and nonmonetary forms.

The private sector's decision to recycle will vary according to the particular waste stream, according to the geographical location of the waste stream (for example, the 1984 cost of landfill varied from $1.41 to $29.22 per ton in 1981 dollars depending on the location), and according to which private-sector groups must be involved in the recycling activity. Recall that the private sector can be divided into four main groups — consumers of plastic products, manufacturers in their role as producers of products containing plastics and in their role as producers of manufacturing wastes, waste processors, and recyclers of products (such as automobiles) that contain plastics that are normally disposed as a by-product of the recycling operation. Each group plays a specific role in the recycling of plastics. Further, each group faces its own specific incentives and barriers to recycle; and whether a particular party is involved in the decision to recycle will largely depend on the economic and institutional parameters that are relevant to a particular waste stream.

The consumer's main role in plastics recycling is diverting plastics from the municipal waste stream. Manufacturers can contribute to recycling by labeling and designing plastic products to facilitate their identification and processing during recycling. The remaining private-sector groups are more directly involved in the decision to adopt a specific recycling technology or dispose of the waste by landfill or incineration.

In the case of consumers, incentives to divert plastic wastes from the municipal waste stream may include environmental concerns about plastics disposal; direct incentives to retrieve a deposit on a plastic product, such as a beverage bottle; or an indirect incentive to obtain the scrap value of products that contain plastic parts. Barriers include the cost and inconvenience of source separation and the cost of transportation to a collection point.

In the case of manufacturers of products that contain plastic components, incentives to design and build products that are more easily recycled include goodwill toward potential customers and government regulators. They also include a desire to make their products more valuable once scrapped, which, in the case of products normally diverted from the municipal waste stream, reduces the effective cost of their products to consumers. A barrier to the manufacturer is the potential additional cost of designing and building products that are more easily recycled.

Manufacturers as producers of manufacturing waste, waste processors, and recyclers of products that contain plastics as by-products have the main incentive of avoiding the cost of disposal, in addition to the potential goodwill that a recycling operation may generate. Barriers are mostly in the form of technological, market, and regulatory uncertainties that are posed by the adoption of a recycling operation. A major conclusion from Chapter 3 is that, while the incentives and barriers faced by different private-sector groups vary widely, those incentives and barriers are often interdependent.

Evidence to date indicates that successful recycling operations have taken advantage of institutional situations that largely eliminate many of the potential barriers to recycling. Chapter 6 has discussed several examples where plastic wastes are diverted from the municipal waste stream by government intervention or because of long-standing institutional arrangements. Technological uncertainties have been overcome by adopting equipment that is often a slight variation of conventionally used equipment. Market uncertainties have often been minimized in that the recycler is also the consumer of the recycled goods. In the coming decade it is likely that the recycling of plastics by the private sector will continue to utilize technologies with minimal technical risks and occur in market segments where institutional factors have eliminated many of the would-be barriers to recycling.

IS THERE A NEED FOR THE PUBLIC SECTOR TO PROMOTE PLASTICS RECYCLING?

From the perspective of economic efficiency, the degree to which the public sector is needed to promote plastics recycling depends on two factors: market failures that prevent the private sector from providing the most efficient level of recycling, and market distortions that result from government regulations and legislation. Chapter 3 of this book discusses the various market failures and government regulations and legislation that may have an impact on the private sector's recycling activities. In this concluding chapter let us focus on the key parameters around which the public's decision must be made.

The most obvious market failure that may result in the private sector's not providing the most efficient level of plastics recycling is the existence of externalities, i.e., costs or benefits that result from the production, use, or disposal of a good or service, but that are not borne by the parties making the economic transaction. Because these costs and benefits are not a relevant consideration to the buyer and seller of the good or service, they are not included in the purchase price. In the case of positive externalities, the good or service will be priced too high and thus underused from the perspective of economic efficiency. In the case of negative externalities, the good or service will be priced too low and thus overused. In both cases the government can intervene to force the price of the good or service to reflect its true costs and benefits, inclusive of the costs and benefits borne by individuals not directly involved in the transactions, i.e., the general public, in the case of plastics recycling. A more efficient allocation of total resources will therefore result.

The main externality to be considered by the public sector is the potential environmental damage that may occur from the disposal of plastic wastes. Unfortunately, as discussed earlier in this chapter and in more detail in Chapter 2, the answer to this question is controversial. Any position is both technically defensible and vulnerable, depending on the technical expert consulted. Without additional information on the environmental impacts of plastics disposal, the public's decision about this very important externality will continue to be based on political considerations, rather than the more defensible criterion of economic efficiency.

Given the current uncertainty about the environmental implications of plastics disposal, the more defensible reason for government promotion of recycling is the current way most household wastes are disposed of. Households do not normally pay the marginal cost of waste disposal. In other words, households typically pay a flat fee for trash disposal, implying that the additional cost of an additional pound of waste, whether it be a plastic or otherwise, is zero. Households therefore have no incentive to source separate their waste or to promote recycling, other than their own personal preferences for recycling over disposal or their desire to retrieve a deposit, such as those on plastic bottles in many states. This institutional arrangement is desirable in the sense that it does not encourage households to dispose of their wastes outside of the municipal waste stream so they can avoid a positive marginal cost of disposal. However, the arrangement also implies that most households will not give much consideration to the cost of disposal — including the external costs — when purchasing or disposing a good. Government intervention in some form may therefore be desirable to help remove the inherent advantage that disposal has over recycling. However, this intervention is not applicable only to plastics. The same argument can be used for any material entering the waste stream.

Other externality arguments can be made on less defensible grounds. For example, it can be argued that because most plastics contain heat values equivalent to coal, landfill or incineration of plastics without heat recovery will impose an externality in that this heat content is lost. To the extent that the tertiary or quaternary use of plastic wastes could reduce the importation of oil, it can be argued that the now well established concept of the "oil import premium" applies, and plastics recycling in this form should be subsidized.

(See Chapter 3 for details.) However, once again, this argument does not apply only to plastic wastes. It applies to all forms of domestic energy and gives no preference to the potential energy from plastics recycling.

The other set of arguments for government assistance in plastics recycling is based on current government intervention that is claimed to interfere with the normal market process. These claims include, for example, tax inequities and freight-rate differences between virgin and recycled materials; zoning laws that force recycling centers to locate far from sources of waste; and regulations that require products containing recycled materials to be labeled as such, which may unfairly bias the consumer's perception of the goods. To some extent these arguments apply to plastics recycling. However, as is discussed in detail in Chapter 3, the arguments are often difficult to follow because, for example, it is difficult to identify the markets in which recycled plastic products might compete. Secondary plastic products will probably compete with products normally made from wood, metals, or concrete. Therefore, should tax and freight rates for plastic wastes and recycled plastic products be equal to those paid for other materials? Further, in the case of freight rates, it is difficult to argue that the differences reflect anything other than actual cost differences.

To the extent that political and other constraints prevent the removal of the regulatory and legal impediments to plastics recycling — which is the first best alternative — government assistance to promote recycling is appropriate. However, once again, these types of arguments can be made with respect to most any material that enters the waste stream. They do not give a preference to plastics recycling over the recycling of other materials.

Therefore, while it can successfully be argued that the government has a role in plastics recycling, it is not currently clear what the proper extent of that role should be. Additional research is warranted to determine the extent to which the private sector's response to the problems will differ from the optimal social response. The major consideration here is the current uncertainty about the environmental impacts of plastics disposal and recycling.

HOW CAN GOVERNMENT PROMOTE PLASTICS RECYCLING?

In general, the federal government can promote plastics recycling in two ways: a) by encouraging local and state governments to recycle rather than dispose of their municipal solid waste, which typically contains about six percent plastics; and b) by encouraging the private sector to divert plastics from, and recycle those plastics outside of, the municipal waste stream. At a more basic level, government can promote recycling by increasing the incentives and reducing the barriers to recycling as perceived by those involved in that decision.

Given that it will be economically, institutionally, and technically difficult to divert more than about 25 percent of all plastic wastes from the municipal waste stream in the coming decade, government can promote the recycling of the greatest quantity of plastics by promoting the tertiary and quaternary recycling of municipal waste. Municipalities can be encouraged to recycle using two basic approaches. On the one hand, disposal can be made more costly by imposing and enforcing stricter regulations on the currently accepted means of disposal. On the other hand, recycling can be made relatively less costly by providing government funds to develop less costly tertiary and quaternary processes or by providing direct subsidies to recycling operations. A decision about the degree to which recycling should be promoted depends on the direct and external costs of disposal versus recycling. And while the direct costs are fairly well documented, the external costs remain a point of contention. Obviously, the recycling of plastics is only one part of a decision about recycling municipal wastes. The recycling of other materials in the waste stream must complement the recycling of plastics.

The more complicated part of the government intervention question concerns how the government through its programs should alter the private sector's perspective of recycling. Government actions that impact different private-sector decision makers are crucial in determining whether the approximately 25 percent of plastics waste that can be diverted from the municipal waste stream will actually be diverted. Without this diversion, secondary recycling is not feasible, as well as many tertiary and quaternary processes that require relatively clean plastics.

Whether the diversion of plastic wastes from the municipal waste stream is desirable from the social perspective depends on the external cost of the wastes to society when processed in, and out of, that stream. As stated earlier, the answer to this question is controversial. However, if we assume that the diversion of plastic wastes from the municipal waste stream is desirable, several government policies are available. Consumers can be encouraged to divert wastes from the municipal waste stream by deposit laws, such as the current bottle laws, by providing collection services for segregated recyclable materials, as stipulated by the recently enacted Oregon Recycling Opportunity Act; by requiring producers of plastic products to design and label those goods to facilitate the separation of different plastic types; and by providing information to the public about the ways plastics can be recycled. Manufacturers of plastic wastes, waste processors, and recyclers of products that contain plastics as by-products can be encouraged to recycle by placing a tax on disposal (with the unfortunate side-effect of promoting the open dumping of waste); by reducing the technological uncertainties about plastics recycling by providing government support for plastics recycling R&D; by reducing market uncertainties through government procurement of recycled products, and by providing information about available recycling processes, applicable waste streams, and potential customers of recycled goods; and by reducing regulatory uncertainties by establishing consistent and nonfluctuating government policies toward waste disposal and recycling operations. However, with each of these potential measures, government must be careful not to impose costs on society that exceed the costs that society would bear if the waste is simply disposed of by conventional means. It must be recognized that government promotion of recycling will in most cases impose its own social costs. If efficiency is to be served, those costs must not exceed the external costs society would pay in the event that the private sector is left to decide the fate of plastics recycling.

IS PLASTICS RECYCLING AN ECONOMICALLY VIABLE ALTERNATIVE TO DISPOSAL?

Chapter 5 of this book reviews several published estimates of the direct costs and revenues associated with various forms of plastics recycling. Also included are estimates of the direct costs of landfill and incineration without heat recovery. Two main conclusions can be drawn from that review. First, for those plastic wastes that can be segregated from other waste materials and are applicable to the recycling technologies for which cost and revenue estimates are available, recycling is generally a more attractive alternative than either disposal or recycling within the municipal waste stream. However, the current data will not allow a distinction to be made among the various secondary, tertiary, and quaternary processes. Second, for those wastes that enter the municipal waste stream, the direct costs of disposal will generally be less than the direct costs of recycling. Landfill is usually cheaper than incineration. However, depending on the geographical location and population size, recycling by quaternary methods may surpass disposal. The tertiary recycling of municipal waste in the form of pyrolysis is not currently being used.

A major emphasis of this book, however, is that when judging the economic feasibility of recycling plastic wastes, one cannot simply compare the direct costs of recycling with the alternative of disposal. In many cases technical constraints will limit the applicability of a waste stream to a particular recycling technology. And what may be equally important, the constraints and subsequent monetary and nonmonetary costs that are imposed on recycling by numerous institutional barriers may render an otherwise economically viable recycling operation infeasible.

In the coming decade the most likely scenario is that the private sector will continue to adopt plastics recycling as an alternative to disposal in those limited cases where plastics can be easily collected outside of the municipal waste stream. Recycling technologies that are variations of conventional equipment will tend to be used because of their minimal technological risks. In addition, plastic waste streams that pose minimal institutional barriers and provide maximum institutional benefits will receive preference. If the historical adoption of recycling indicates future adoption patterns, the economic feasibility of plastics recycling will be determined as much by institutional factors as by technological considerations.

From the public perspective, the economic feasibility of recycling must be based not only on the direct costs and revenues considered by the private sector. External costs, mostly in the form of environmental degradation, will largely determine if the private-sector's response to plastics recycling is acceptable or unacceptable. Until a consensus is reached on the extent of these externalities, the appropriate public-sector response to plastics recycling will continue to be the subject of significant debate.

Appendix A

Glossary of Characteristics and Uses of Selected Plastic Resins

THERMOPLASTIC RESINS

Thermoplastic resins consist of either linear or branched polymer molecules that are not attached to other polymer molecules. This characteristic allows thermoplastics to be repeatedly softened and hardened by heating and cooling, with few chemical changes taking place during the process. Thermoplastic waste can therefore be melted under heat and pressure, reformed while in a molten or softened state, and cooled so that the recycled plastic hardens to the desired shape. Thermoplastics include the following major resins.

ABS (Acrylonitrile-butadiene-styrene) possesses outstanding impact and mechanical strength. ABS exhibits good dimensional stability and electrical insulating properties. Typical uses include automotive parts (such as automotive trim and grilles, which can be coated with a chrome-like metallic finish), electrical housings (such as telephone bodies), refrigerator liners, construction products (such as bath tubs), and sporting goods. ABS is an impact-modified version of SAN.

Nylons are a large family of resins characterized by high tensile strength, good impact strength, abrasion resistance, resistance to moderate heat, and good chemical resistance and electrical properties. Typical uses include brush bristles, wire jackets, fender extensions, fishing line, cooking bags, fibers, and gears and other sliding-contact products.

Polyesters (Thermoplastics) are characterized by high melting point (in excess of 400°F), high resistance to abrasion, good chemical resistance, hardness, and

149

low moisture absorption. Applications include gears, bearings, pulleys, furniture, fender extensions, and other products that might otherwise require metals or equally strong materials. Thermoplastic polyesters are also used for some packaging applications, such as carbonated beverage bottles.

Polyethylene is characterized by toughness, excellent chemical resistance, excellent electrical insulating properties, and near impermeability to moisture. Polyethylene is categorized into low-density and high-density types. Low-density polyethylene (LDPE) is flexible, has high impact strength, and low resistance to heat (under 200°F), which facilitates processing. Typical uses of LDPE are household cling wrap, housewares, squeeze bottles, pipes, and toys. High-density polyethylene (HDPE) exhibits greater stiffness and rigidity, improved heat resistance, and increased resistance to permeability when compared with LDPE. Applications include seating; large shipping containers and drums; garbage cans; gasoline tanks; grocery bags; and bottles for milk, chemicals, and detergents.

Polypropylene has characteristics including low density, excellent chemical resistance, negligible water absorption, excellent electrical insulation, and low resistance to heat, which makes processing easier. Uses for polypropylene include automotive battery cases; carpet backing; bottles for foods, such as syrups; automotive fender skirts; toys; housewares; and uses in some household appliances, such as microwave ovens and washing machines. Polypropylene can also be made into foams and molded into furniture frames.

Polystyrene comprises a large family of resins that exhibit varying characteristics. At one extreme are relatively cheap styrenes that exhibit good hardness and rigidity. At the other extreme are more expensive styrenes that exhibit excellent impact resistance, rigidity, and toughness. Polystyrene is also available as liquid solutions and adhesives, and can be made into foams. Polystyrene foams have excellent low-thermal conductivity, low water absorption, good chemical resistance, and a high strength-to-weight ratio. Low-end polystyrene uses include disposable serviceware and toys. The more expensive and tougher styrenes are used in business-machine housings, automotive parts, appliance housings, wall tiles, and video cassettes. Polystyrene foams are used as insulation board for construction, egg cartons, insulated drink cups, marine applications, and housewares.

PVAc (Polyvinyl Acetate) and other vinyls comprise a wide variety of vinyls that exhibit varying properties. They can be formulated to have good mechanical properties, abrasion resistance, or water solubility. Uses include adhesives, paints, coatings, and packaging. Vinyl chloride/vinyl acetate copolymer resins are used in phonograph records.

PVC (Polyvinyl Chloride) possesses excellent chemical and weather resistance, abrasion resistance, and a slow rate of water absorption. Low resistance to heat (less than 200°F) facilitates processing. Applications include food wrap, flooring, upholstery, wire insulation, pipe, shower curtains, and garden hose.

SAN (Styrene-Acrylonitrile) has characteristics including rigidity, transparency, and outstanding chemical resistance. Typical uses include automotive trim and instrument panels, housewares, solar collectors, and boat hulls. Also see ABS.

THERMOSETTING RESINS

Thermosetting resins consist of molecular structures that are cross-linked into a three-dimensional structure during polymerization. This cross-linking characteristic prevents thermosets from being softened by heating without degrading some of the linkages. Therefore thermosetting resins cannot be heated and reformed into new products as can the thermoplastic resins. This characteristic severely limits the recycling possibilities for thermoset wastes. Thermosets include the following major resins.

Epoxy has characteristics that include high strength, low moisture absorption, excellent electrical properties, and excellent adhesive characteristics for bonding metals, ceramics, plastics, glass, and other materials. Epoxies are often used in combination with glass fibers to form high-strength composites for use in aircraft components, pipes, and pressure tanks. Epoxies are also used as adhesives, protective coatings in appliances, industrial equipment, sealants, and in the encapsulation and lamination of various electrical and electronic components.

Melamine's properties include extreme hardness, excellent colorability, and good chemical resistance. Uses include dinnerware, buttons, handles, and knobs. In liquid form, melamines are used as adhesives, coatings, and laminates. Counter tops are usually made with melamine in combination with papers. Chemically, melamine is closely related to urea.

Phenolic's characteristics include good heat resistance, good mechanical properties, good electrical properties, and ease of moldability. Uses include automotive parts (such as distributor caps, transmission parts, and power-brake components), electrical and electronic connectors and switches, and appliance handles and knobs. Phenolic is also used as an adhesive and a bonding agent for abrasives, insulation, and laminates. Another use for phenolic is in binding together the various layers of wood in plywood.

Polyester (Unsaturated) has characteristics that vary significantly depending on the specific formulation. Some thermosetting polyesters are soft and flexible, while others are hard and brittle. These resins are usually used in combination with reinforcements, such as glass fibers, to form composite structures. Typical uses include automotive components, construction goods (such as showerstalls, synthetic marble, and sanitary fixtures), and consumer goods (such as luggage, fishing poles, and bowling balls).

Urea-formaldehyde resins have characteristics including extreme hardness, scratch resistance, good chemical resistance, and good electrical properties. Uses include lighting fixtures, buttons, wiring devices, toilet seats, and decorative housings. In liquid form, urea is used as an adhesive and bonding agent. Chemically, urea is closely related to melamine.

POLYURETHANE FOAMS

Polyurethane foams are a class of resins that are usually considered to be thermosets, but often have characteristics similar to thermoplastics. Polyurethane foams are therefore often discussed separately from other thermosets and thermoplastics. Polyurethane foams have excellent thermal insulating properties, and, depending on the additives used, can be either hard, soft, flexible, or tough. Flexible foams have excellent energy absorbing characteristics and are often used in furniture cushioning, carpet underlay, bedding, packaging, and automotive seating and safety padding. Rigid foams have excellent heat insulation, dimensional stability, and flotation characteristics and are used in insulations for buildings and refrigerators, flotation equipment for boats, and structural and nonstructural components in furniture.

Appendix B

Plastic Waste Projections: Methodology and Detailed Results

INTRODUCTION

This appendix presents a detailed discussion of the methodologies used to obtain the estimates and projections discussed in Chapter 4. Also presented are numerical results that are presented in graphic form in Chapter 4.

HISTORICAL AND PROJECTED U.S. RESIN PRODUCTION AND USE

Methodology

The total production of plastic resins in the United States was projected through the year 1995. Also projected were the quantities of resins used in nine major U.S. product categories — packaging, consumer and institutional goods, transportation, electrical and electronic goods, construction, furniture and fixtures, industrial equipment, adhesives, and other goods. Projections were also obtained for U.S. exports of plastic resins.

Given that the supply and demand for plastics as a whole and as inputs to particular products have undergone significant structural change during recent years, a relatively simple methodology was adopted to obtain projections of U.S. resin production and use. All projections were made using time series analysis. However, in the interest of studying the historical relationships between resin production and use and the level of production in certain product categories, the regressions included an appropriate industrial production index as an independent variable. Included in this appendix are projections of future values of these production indexes, which were derived using time series analysis. Note that the inclusion of these production indexes in the regression equations does not alter the projections of resin production and use, since future values of the indexes are based solely on time. The method of ordinary least squares was used for all regressions.

All data on the production and use of plastic resins were obtained from various issues of *Facts and Figures of the U.S. Plastics Industry* published by the Society of the Plastics Industry. Production index data were obtained from various issues of the *Survey of Current Business* published by the U.S. Bureau of Economic Analysis within the U.S. Department of Commerce.

Table B.1 defines the variable names used in the remainder of this appendix. All variables represent U.S. production or use.

Table B.2 gives details of the regressions used to project values for the production indexes. Table B.3 gives historical data and projected values for the production indexes used in the projections of resin production and use. Historical data are given for the years 1973–84. Projected values are given for the years 1985–95.

Results

The results of the regressions used in the projections of total resin production and the use of resins in selected product categories are summarized in Table B.4. Note that in most cases a reasonably good fit is obtained. As we would expect, the coefficient on the production index used in all regressions is positive and is significant in all but one case. The coefficient on time is either positive or negative depending on the specific equations, reflecting either an increase or decrease in the use of resins in a particular product category over time when adjusted for variations in the output of that product category. Note that in most cases where time has a negative coefficient, the coefficient is not highly significant.

Table B.5 presents data on the historical and projected production of resins in the U.S. and the consumption of resins in specific domestic product categories. Historical data are given for the years 1974–83 in the case of total resin production and for 1974–84 in all other cases. All other data entries are projected values derived from the methodology discussed above.

The historical data given in *Facts and Figures of the U.S. Plastics Industry* on the use of resins in major product categories do not include the use of polyurethane. The use of polyurethane is provided separately in that publication. In keeping with that distinction, this work projected the use of polyurethane separately from other resins. The methodology used to obtain the projections for polyurethane is the same as used for other resins. Table B.6 summarizes the regression results used to make the projections for polyurethane in the product categories that use polyurethane. The fit of the regressions varies from very good in the case of building and construction materials to very poor in the case of electrical and electronic goods.

TABLE B.1. Definition of Variables

Variable Name	Definition of Variable
RESPRO	Total resin production
RESPAK*	Total resins used in packaging
RESC&I*	Total resins used in consumer and institutional goods
RESTRA*	Total resins used in the transportation sector
RESE&E*	Total resins used in the electrical and electronics sector
RESBUI*	Total resins used in the building and construction sector
RESFUR*	Total resins used in furniture and fixtures
RESIND*	Total resins used in industrial equipment
RESADH*	Total resins used in adhesives
RESEXP*	Total resins exported
RESOTH*	Total resins used in goods not included in other product categories
PFTRAN	Polyurethane used in the transportion sector
PFPACK	Polyurethane used in packaging
PFBUIL	Polyurethane used in building and construction
PFELEC	Polyurethane used in electrical and electronic goods
PFFUR	Polyurethane used in furniture and fixtures
PFOTH	Polyurethane used in products not included in other product categories
PITOT	Production index for all goods
PIMAN	Production index for all manufacturing goods
PINON	Production index for all nondurable goods
PIDUR	Production index for all durable goods
PIC&I	Production index for consumer goods
PIINDU	Production index for industrial equipment
PIFUR	Production index for furniture and fixtures
PITRAN	Production index for transportation equipment
PICONS	Production index for construction supplies
PIELEC	Production index for electrical machinery

Note: *Excludes the use of polyurethane

Table B.7 gives the historical and projected use of polyurethane in the market categories where that use is concentrated. The data entries for the years 1973–83 reflect historical usage. The remaining entries are projections.

PROJECTIONS OF MANUFACTURING NUISANCE PLASTICS

Methodology

A detailed discussion of the methodology used to project future quantities of manufacturing nuisance plastics can be found in Chapter 4.

TABLE B.2. Summary of Regression Results: Production Indexes

Dependent Variable	R-squared	Time Interval	Constant	Coefficient		
				t-stat	Time	t-stat
PITOT	0.70	1973–1984	− 6,052.27	− 4.68	3.13	4.79
PIMAN	0.70	1973–1984	− 6,369.75	− 4.76	3.29	4.87
PINON	0.88	1973–1984	− 8,624.03	− 8.23	4.44	8.38
PIDUR	0.50	1973–1984	− 4,800.76	− 3.08	2.49	3.16
PIC&I	0.74	1973–1984	− 4,930.90	− 5.21	2.56	5.36
PIINDU	0.16	1973–1984	− 3,294.48	− 1.35	1.73	1.40
PIFUR	0.82	1973–1984	− 10,155.04	− 6.72	5.21	6.82
PITRAN	0.29	1973–1984	− 3,847.75	− 1.95	2.00	2.01
PICONS	0.23	1973–1984	− 3,317.43	− 1.67	1.75	1.74
PIELEC	0.88	1973–1984	− 15,448.75	− 8.61	7.89	8.69

Additional Results

In addition to the results presented in Chapter 4, this section gives estimates and projections of manufacturing nuisance plastics disaggregated by specific resin for the years 1984, 1990, and 1995. Table B.8 gives the results of that disaggregation. The total quantities of nuisance plastics (discussed in Chapter 4) are disaggregated according to the percentage composition of total 1983 plastic resins produced in the United States. No information was found that identifies which of the resins are more prone to become manufacturing nuisance plastics, thus the use of this simplifying assumption.

Phenolics are projected to be the most produced thermosetting manufacturing nuisance plastic at 6.0 percent of the total. With respect to thermoplastic resins, LDPE and HDPE combine to give 32.5 percent of the total projected manufacturing waste, followed by PVC at 14.5 percent and polypropylene at 10.5 percent. Polyurethane foams are projected to compose 4.5 percent of the total.

POSTCONSUMER PLASTIC WASTES

Methodology

The general methodology to project postconsumer plastic wastes outlined in Chapter 4 varies somewhat by product category because of limited data availability and varying life spans. This section gives a detailed discussion of the methodology used to project postconsumer wastes in each of the eight

TABLE B.3. Historical and Projected Values for Selected Production Indexes

obs	PITOT	PIMAN	PIDUR	PIC&I	PIINDU	PIFUR	PITRAN	PICONS	PIELEC
1973	125.6000	125.1000	122.0000	131.7000	120.1000	126.1000	109.1000	133.8000	126.8000
1974	124.8000	124.4000	120.7000	128.8000	128.7000	126.9000	96.90000	129.6000	125.2000
1975	117.8000	116.3000	109.3000	124.0000	121.2000	118.2000	97.40000	116.3000	116.5000
1976	129.8000	129.5000	121.7000	136.2000	128.0000	132.7000	110.6000	132.6000	131.6000
1977	137.1000	137.1000	129.5000	143.4000	138.5000	140.9000	121.1000	140.8000	141.9000
1978	145.2000	145.6000	139.3000	147.4000	149.9000	154.7000	130.5000	153.3000	154.3000
1979	152.5000	153.6000	146.4000	150.8000	152.2000	161.5000	135.4000	158.0000	175.0000
1980	147.2000	146.6000	136.6000	145.5000	156.9000	149.8000	116.6000	140.9000	172.7000
1981	151.0000	150.4000	140.5000	147.9000	166.4000	157.2000	116.1000	141.9000	178.4000
1982	136.6000	137.6000	124.7000	142.6000	134.9000	151.9000	104.9000	124.3000	169.3000
1983	147.6000	148.2000	134.5000	151.7000	120.4000	170.5000	117.8000	142.5000	185.5000
1984	157.0886	164.8000	154.6000	161.6000	140.6000	190.2000	137.6000	158.9000	217.4000
1985	160.2183	161.3198	147.8548	159.3017	149.4272	182.2329	129.1895	150.7652	209.1562
1986	163.3480	164.6101	150.3478	161.8661	151.1622	187.4406	131.1930	152.5124	217.0444
1987	166.4778	167.9003	152.8408	164.4304	152.8972	192.6483	133.1965	154.2596	224.9325
1988	169.6075	171.1905	155.3338	166.9947	154.6321	197.8560	135.2000	156.0068	232.8206
1989	172.7372	174.4807	157.8268	169.5591	156.3671	203.0637	137.2035	157.7540	240.7087
1990	175.8669	177.7709	160.3198	172.1234	158.1021	208.2714	139.2070	159.5012	248.5968
1991	178.9966	181.0611	162.8128	174.6877	159.8370	213.4791	141.2105	161.2484	256.4849
1992	182.1264	184.3513	165.3058	177.2521	161.5720	218.6868	143.2140	162.9956	264.3730
1993	185.2561	187.6415	167.7988	179.8164	163.3070	223.8945	145.2175	164.7428	272.2611
1994	188.3858	190.9317	170.2918	182.3807	165.0419	229.1022	147.2210	166.4900	280.1492
1995	191.5155	194.2219	172.7848	184.9451	166.7769	234.3099	149.2245	168.2372	288.0374

157

TABLE B.4. Summary of Regression Results: Resin Production and Use

Dependent Variable	R-squared	Time Interval	Coefficient				Production Index	Coefficient	t-stat
			Constant	t-stat	Time	t-stat			
RESPRO	0.94	1973–1983	−823,919.22	−1.62	405.06	1.55	PITOT	415.78	5.70
RESPAK	0.97	1974–1984	−826,014.07	−6.58	417.79	6.49	PIMAN	60.09	3.94
RESC&I	0.81	1974–1984	−19,815.51	−0.29	9.57	0.28	PIC&I	29.96	2.78
RESTRA	0.68	1974–1984	38,494.12	1.02	−19.16	−1.02	PITRAN	17.76	3.89
RESE&E	0.79	1974–1984	119,691.23	2.62	−61.29	−2.62	PIDUR	31.47	5.25
RESBUI	0.97	1974–1984	−800,845.46	−11.56	404.27	11.46	PICONS	52.60	6.10
RESFUR	0.69	1974–1984	92,683.10	1.47	−47.18	−1.46	PIFUR	15.95	3.05
RESIND	0.31	1974–1984	21,388.39	1.40	−10.81	−1.39	PIINDU	2.78	1.62
RESADH	0.77	1974–1984	368,493.76	5.06	−188.16	−5.05	PIMAN	43.52	4.93
RESOTH	0.95	1974–1984	−209,067.69	−3.94	105.07	3.86	PITOT	29.01	4.01
RESEXP	0.90	1974–1984	−505,963.73	−4.34	256.60	4.30	PIMAN	7.59	0.54

Note: All regressions exclude the use of polyurethane

TABLE B.5. Historical and Projected Production and Use of Plastics (in millions of pounds)

obs	RESPRO	RESPAK	RESC&I	RESTRA	RESE&E	RESBUI	RESFUR	RESIND	RESOTH	RESADH	RESEXP
1974	29274.00	6720.000	3168.000	1725.000	2524.000	4327.000	1791.000	488.0000	2215.000	2150.000	1585.000
1975	22828.00	5579.000	2875.000	1248.000	1787.000	3736.000	1360.000	313.0000	1771.000	1942.000	1351.000
1976	29196.00	7342.000	2801.000	1808.000	2524.000	4555.000	1617.000	319.0000	2160.000	2330.000	2168.000
1977	33948.00	7899.000	3242.000	1911.000	2756.000	6008.000	1391.000	472.0000	2728.000	2566.000	2142.000
1978	37605.00	9044.000	3592.000	2015.000	2952.000	6965.000	1686.000	380.0000	2939.000	2902.000	2588.000
1979	41577.00	10334.00	3753.000	1934.000	3043.000	7573.000	1894.000	517.0000	3443.000	2794.000	3432.000
1980	37347.00	10003.00	3553.000	1605.000	2453.000	6424.000	1646.000	391.0000	3054.000	2387.000	3670.000
1981	39867.00	10465.00	3670.000	1573.000	2670.000	7259.000	1670.000	393.0000	3259.000	2572.000	3425.000
1982	36607.00	10497.00	3269.000	1392.000	2275.000	7154.000	1556.000	241.0000	3232.000	1584.000	3909.000
1983	42777.00	11813.00	3816.000	1896.000	2514.000	8552.000	2007.000	318.0000	3636.000	1800.000	4150.000
1984	45039.93	12398.00	3986.000	2109.000	2757.000	9691.000	2117.000	400.0000	4080.000	1993.000	3796.000
1985	46746.25	12992.15	3956.642	1849.115	2683.247	9563.839	1927.552	346.6183	4156.739	2017.665	4620.450
1986	48452.58	13607.65	4043.051	1865.085	2700.412	10060.01	1963.412	340.6246	4352.816	1972.702	4902.027
1987	50158.90	14223.16	4129.460	1881.056	2717.577	10556.19	1999.272	334.6308	4548.892	1927.738	5183.603
1988	51865.22	14838.66	4215.869	1897.026	2734.741	11052.36	2035.132	328.6369	4744.968	1882.774	5465.180
1989	53571.54	15454.17	4302.278	1912.997	2751.906	11548.54	2070.992	322.6432	4941.045	1837.810	5746.757
1990	55277.87	16069.67	4388.687	1928.967	2769.070	12044.71	2106.852	316.6494	5137.122	1792.846	6028.333
1991	56984.20	16685.18	4475.096	1944.937	2786.235	12540.89	2142.711	310.6556	5333.198	1747.883	6309.909
1992	58690.52	17300.69	4561.505	1960.908	2803.399	13037.06	2178.571	304.6618	5529.275	1702.919	6591.486
1993	60396.85	17916.19	4647.914	1976.878	2820.564	13533.24	2214.431	298.6680	5725.351	1657.955	6873.062
1994	62103.17	18531.70	4734.323	1992.849	2837.729	14029.41	2250.291	292.6742	5921.427	1612.991	7154.639
1995	63809.49	19147.20	4820.732	2008.819	2854.893	14525.58	2286.151	286.6804	6117.504	1568.027	7436.215

Source: Historical data from Society of the Plastics Industry, *Facts and Figures of the U.S. Plastics Industry* (various issues). Used with permission.

TABLE B.6. Summary of Regression Results: The Use of Polyurethane in Selected Product Categories

Dependent Variable	R-squared	Time Interval	Coefficient				Production Index	Coefficient	t-stat
			Constant	t-stat	Time	t-stat			
PFTRAN	0.85	1973–1983	39,660.20	5.97	−20.16	−5.96	PITRAN	4.97	5.41
PFPACK	0.96	1973–1983	−10,586.90	−6.44	5.34	6.33	PIMAN	0.53	2.36
PFBUIL	0.99	1973–1983	−46,521.36	−34.47	23.51	34.24	PICONS	1.99	10.58
PFELEC	0.07	1973–1983	−844.88	−0.71	0.47	0.77	PIDUR	−0.06	−0.34
PFFUR	0.74	1973–1983	31,368.64	0.88	−16.22	−0.89	PIFUR	10.53	2.92
PFOTH	0.17	1973–1983	−10,170.88	−1.12	5.24	1.12	PITOT	−0.65	−0.50

TABLE B.7. **Historical and Projected Use of Polyurethane in Selected Product Categories (in million of pounds)**

obs	PFTRAN	PFPACK	PFBUIL	PFELEC	PFFUR	PFOTH
1973	410.0000	25.00000	140.0000	75.00000	660.0000	107.0000
1974	367.0000	21.50000	140.0000	77.50000	602.5000	99.00000
1975	324.0000	18.00000	140.0000	80.00000	545.0000	91.00000
1976	390.5000	33.50000	188.0000	80.50000	748.0000	87.00000
1977	457.0000	49.00000	236.0000	81.00000	951.0000	83.00000
1978	466.0000	57.00000	283.0000	79.00000	980.0000	96.00000
1979	418.0000	62.00000	322.0000	84.00000	973.0000	71.00000
1980	288.0000	57.00000	315.0000	69.00000	872.0000	111.0000
1981	268.0000	77.00000	328.0000	76.00000	898.0000	97.00000
1982	248.0000	77.00000	328.0000	80.00000	859.0000	82.00000
1983	301.0000	85.00000	379.0000	87.00000	872.0000	185.0000
1984	353.1400	97.61606	437.9070	80.24431	1197.568	119.6335
1985	291.2125	101.0996	445.1970	81.14203	1097.473	122.8307
1986	281.0064	108.1967	472.1899	81.45562	1136.084	126.0279
1987	270.8003	115.2938	499.1828	81.76921	1174.695	129.2251
1988	260.5942	122.3909	526.1756	82.08281	1213.306	132.4223
1989	250.3882	129.4880	553.1686	82.39640	1251.917	135.6195
1990	240.1821	136.5851	580.1615	82.71000	1290.528	138.8167
1991	229.9760	143.6823	607.1544	83.02358	1329.139	142.0140
1992	219.7699	150.7794	634.1473	83.33718	1367.750	145.2112
1993	209.5638	157.8765	661.1401	83.65078	1406.362	148.4084
1994	199.3577	164.9736	688.1330	83.96436	1444.972	151.6056
1995	189.1517	172.0707	715.1260	84.27796	1483.583	154.8028

Source: Historical data from Society of the Plastics Industry, *Facts and Figures of the U.S. Plastics Industry* (various issues). Used with permission.

product categories. Data on the historical use of plastic resins in specific product categories have been obtained from various issues of SPI's *Facts and Figures of the U.S. Plastics Industry*.

Transportation

Estimates for 1984.   The average life span for transportation equipment is assumed to be 11 years. [Holcomb and Koshy (1984) report that automobiles have a median life span of 10.9 years.] Using the three-year average method discussed in Chapter 4, information is needed on the use of plastics in the transportation sector for the years 1972, 1973, and 1974. Unfortunately, this information was not found. Information was, however, identified for the

TABLE B.8. **Manufacturing Nuisance Plastics Disaggregated by Resin Type (in millions of pounds)**

	1983 Composition of U.S. Resins (%)	1984	1990	1995
Thermosets				
Epoxy	0.8	23	28	32
Polyester	2.5	71	87	100
Urea and Melamine	3.2	90	111	128
Phenolics	6.0	169	208	240
Other Thermosets	0.8	23	28	32
Total Thermosets	13.3	375	460	531
Thermoplastics				
LDPE	19.0	536	658	759
HDPE	13.5	381	467	539
Polypropylene	10.5	296	363	419
ABS, SAN, and OSBP	3.1	87	107	124
Polystyrene	8.5	240	294	340
SBL and OSBL	1.6	45	55	64
Nylon	0.8	23	28	32
PVC	14.5	409	502	579
PVAc and Other Vinyls	2.4	68	83	96
Thermoplastic Polyesters	1.8	51	62	72
Other Thermoplastics	6.5	183	225	260
Total Thermoplastics	82.2	2,319	2,845	3,284
Polyurethane Foam	4.5	127	156	180
Total	100.0	2,821	3,461	3,995

Notes: Low-Density Polyethylene (LDPE); High-Density Polyethylene (HDPE); Acrylonitrile-Butadiene-Styrene (ABS); Syrene-Acrylonitrile (SAN); Other Styrene-Based Polymers (OSBP); Styrene Butadiene Latexes (SBL); Other Styrene-Based Latexes (OSBL); Polyvinyl Chloride (PVC); Polyvinyl Acetate (PVAc)

Source: 1983 U.S. Resin Composition: Sales and Captive Use, as reported in *Facts and Figures of the U.S. Plastics Industry*, The Society of the Plastics Industry (1984). Used with permission.

years 1974, 1975, and 1976 and was averaged to obtain a proxy for the information needed. This average was disaggregated into specific resin types according to the usage of particular resins in the production of automobiles in 1977—the year closest to the year needed for which information was available by resin type. Automobiles consumed 1.4 billion pounds of the total 1.9 billion pounds of plastics used in the transportation sector in 1977.

Projections for 1990. Total resin usage in the transportation sector for the years 1978, 1979, and 1980 was averaged to project the total quantity of plastic wastes flowing from that sector in 1990. These projections of total waste were disaggregated according to the usage of specific resins in transportation equipment in 1977 — the closest year for which disaggregated information was found.

Projections for 1995. Plastic waste estimates were obtained for the year 1995 by averaging the estimated use of plastics in the transportation sector in the years 1983 and 1984 with the projected use of plastics in that sector in 1985. The projected use of plastics in the transportation sector is discussed in Chapter 4 and in an earlier section of this appendix. This average number is disaggregated into specific resins according to the actual usage of resins in the transportation sector in 1983. The projected use of polyurethane foams was considered separately from the use of other resins. Recall from an earlier section of this appendix that data on the use of polyurethane foams in all sectors is reported separately from the use of other resins. See Tables B.5 and B.7 for all forecasted values.

Packaging

The average life of packaging is less than one year, implying that resins used for packaging become waste the same year they are used for that purpose. In the case of resins other than polyurethane foams, the actual usage of plastics for packaging was used for 1984 estimates and projected values were used for 1990 and 1995 waste projections. These totals were disaggregated according to the use of specific resins in packaging in 1983. Polyurethane foams were considered separately from other resins. The forecasted use of polyurethane foams was used for all estimated and forecasted years. See Tables B.5 and B.7 for all forecasted values.

Building and Construction

The building and construction category, at 25 years, has the longest average life span of all categories considered. This implies that 1984 plastic wastes from this sector were from products produced in about 1959. Wastes in 1990 will be from products produced in about 1965, and the 1995 waste stream will be composed of products produced in about 1970. Unfortunately, detailed information on the consumption of plastics in the building and construction sector by resin type are not available for years so far in the past. The best information found was in the SPI's 1977 edition of *Facts and Figures of the Plastics Industry*. That issue (p. 87) gives information about the use of specific resins in building and construction for the time periods 1960–62 and 1972–74.

The 1960–62 data were used for 1984 estimates and 1990 projections. The 1972–74 data were used for 1995 projections.

Electrical and Electronics Goods

Electrical and electronic goods last an average of about 15 years. Estimates for 1984 and projections for 1990 and 1995 therefore require information about plastics consumption in that sector for the years 1969, 1975, and 1980, respectively. Unfortunately, complete information is not available for the years needed. The average usage of resins in this sector in 1974, 1975, and 1976 was used as a proxy for the needed information in the 1984 estimates and 1990 projections. Data on resin consumption for electrical and electronic goods in 1979, 1980, and 1981 were averaged to project plastic wastes for 1995. The estimates and projections were disaggregated by resin type according to the use of particular resins in the sector in 1981.

Furniture

Furniture products are used an average of 10 years. Estimates for 1984 and projections for 1990 and 1995 therefore require information about plastics consumption in that sector for the years 1974, 1980, and 1985, respectively. Three year averages were used to make all estimates and projections. In the case of resins other than polyurethane foams, resin use in furniture manufacture was averaged for the years 1974, 1975, and 1976 to obtain the 1984 estimate of plastic waste for the furniture sector. (Information was not found for 1973.) Data from 1979, 1980, and 1981 were averaged to project total resin waste for 1990. Consumption of resins for furniture manufacture in 1984 was averaged with the projected use of resins in that sector in 1985 and 1986 to obtain 1995 projections. Estimated wastes for 1984 and projected waste for 1990 were disaggregated into specific resins according to the usage of resins in furniture manufacture in 1981. Projections for 1995 were disaggregated according to the resins used in 1983. The use of polyurethane foam in furniture manufacture was considered separately from other resins. The methodology used was identical to that used with other resins except in the case of 1984 estimates. In that case polyurethane foams used in 1973, 1974, and 1975 were averaged to obtain 1984 waste estimates. Projections of the future use of resins in all product categories are given in Tables B.5 and B.7.

Consumer and Institutional Goods

Consumer and institutional goods have an average life of 5 years. Estimates of total plastic waste from this sector in 1984 were obtained by averaging plastics usage in the sector in 1978, 1979, and 1980. Projections for 1990 were obtained by averaging the usage of all resins in 1984 with the projected use of all resins in 1985 and 1986. The projected use of resins can be found in Tables B.5 and B.7. Estimates for 1984 were disaggregated into specific resin types according to the use of resins in the sector in 1981. The 1990 and 1995 projections were disaggregated according to the composition of resins used in consumer and institutional goods in 1983.

Industrial Machinery

The methodology used for the industrial machinery sector is identical to the methodology described for the electrical and electronics sector. See the above discussion.

Adhesives and Other Goods

It has been assumed by the author that adhesives and other goods have an average life of 4 years. Total estimated waste from the combined sectors for 1984 were obtained by averaging the total consumption of resins in the adhesives and other goods categories during the years 1979, 1980, and 1981. Projections for 1990 were derived from the projected usage of resins in the sector in 1985, 1986, and 1987. The same methodology was used to obtain 1995 projections based on the projected use of resins in 1990, 1991, and 1992. Projections of the future use of resins in these product sectors can be found in Tables B.5 and B.7. Note that polyurethane foams do not appear in the adhesives sector, but do appear in other goods. Estimates for 1984 were disaggregated into specific resins according to resin consumption in that sector in 1981. Projections for 1990 and 1995 were disaggregated according to the 1983 usage of particular resins.

Results

This subsection presents detailed results of the 1984 postconsumer waste estimates and the 1990 and 1995 postconsumer waste projections discussed in Chapter 4. Table B.9 gives estimated 1984 postconsumer plastic wastes by major product category and by several selected resin types. Tables B.10 and B.11 give the same information for the 1990 and 1995 postconsumer projections, respectively. In each table the estimated or projected flow of particular resins from each product category and from all categories combined is given in percentage terms. In addition, information is given in each table about the percentage composition of total postconsumer wastes by major product category.

TABLE B.9. **Estimated Postconsumer Plastic Wastes: 1984 (millions of pounds)**

Plastic Resin	Transpor- tation	%	Packaging	%	Building and Construction	%	Electrical and Electronic
Thermosets							
Epoxy	—	—	—	—	—	—	66
Polyester	473	24.2	—	—	46	6.5	47
Urea and Melamine	—	—	50	0.4	—	—	—
Phenolics	47	2.4	—	—	91	12.8	160
Other Thermosets	65[a]	3.3	12	0.1	7	1.0	59
Total Thermosets	584	29.9	62	0.5	144	20.3	332
Thermoplastics							
LDPE	70[b]	3.6	5,641	45.1	72[b]	10.1	339
HDPE	—	—	2,715	21.7	—	—	111
Polypropylene	332	17.0	893	7.1	—	—	219
ABS and SAN	248	12.7	50	0.4	3	0.4	224
Polystyrene	—	—	1,562	12.5	41	5.8	278
Nylon	59	3.0	37	0.3	—	—	54
PVC	236	12.1	570	4.6	330	46.5	349
PVAc and Other Vinyls	—	—	—	—	—	—	—
Thermopolyester	—	—	595	4.8	—	—	—
Other Thermoplas- tics	65[a]	3.3	260	2.1	119[a]	16.8	372
Total Thermoplastics	1,010	51.7	12,336	98.7	565	79.6	1,947
Total Without Poly- urethane Foam	1,682	81.6	12,398	99.2	709	99.9	2,278
Polyurethane Foam	356[c]	18.2	98	0.8	1[c]	0.1	79
Total	1,954	100	12,496	100	710	100	2,357
Percent	6.7	—	42.2	—	2.4	—	8.0

[a]Data do not distinguish between other thermosets and other thermoplastics.
[b]Data do not distinguish between LDPE and HDPE.
[c]Data do not distinguish between polyurethane foam and other polyurethane uses.
Notes: Low Density Polyethylene (LDPE); High Density Polyethylene (HDPE); Acrylonitrile-Butadine-Styrene (ABS); Syrene-Acrylonitrile (SAN); Polyvinyl Chloride (PVC); Polyvinyl Acetate (PVAc).

%	Furniture	%	Consumer and Institutional	%	Industrial Machinery	%	Adhesives and Other	%	Total	%
2.8	–	–	–	–	–	–	175	3.0	241	0.8
2.0	22	1.0	84	2.3	–	–	93	1.6	765	2.6
–	171	7.8	47	1.3	–	–	175	3.0	443	1.5
6.8	24	1.1	–	–	25	6.9	216	3.6	563	1.9
2.5	–	–	44	1.2	56	15.5	403	6.8	646	2.2
14.1	218	9.9	174	4.8	81	22.4	1,068	18.0	2,663	9.0
14.4	14	0.6	592	16.3	–	–	659	11.1	7,387	24.9
4.7	–	–	476	13.1	94	26.0	560	9.4	3,956	13.4
9.3	774	35.3	483	13.3	51	14.2	490	8.3	3,242	10.9
9.5	–	–	163	4.5	–	–	134	2.3	822	2.8
11.8	59	2.7	748	20.6	–	–	484	8.2	3,172	10.7
2.3	8	0.4	18	0.5	24	6.6	29	0.5	229	0.8
14.8	160	7.3	429	11.8	17	4.8	467	7.9	2,558	8.6
–	38	1.7	76	2.1	–	–	624	10.5	738	2.5
–	–	–	11	0.3	–	–	–	–	606	2.0
15.8	315	14.4	461	12.7	94	26.0	1,325	22.3	3,011	10.2
82.6	1,371	62.6	3,457	95.2	279	77.6	4,768	80.4	25,784	87.0
96.7	1,589	72.5	3,633	100	360	100	5,836	98.4	28,485	96.1
3.3	602	27.5	–	–	–	–	93	1.6	1,229	4.1
100	2,191	100	3,633	100	360	100	5,929	100	29,630	100
–	7.4	–	12.3	–	1.2	–	20.0	–	100	–

TABLE B.10. Projected Postconsumer Plastic Wastes: 1990 (millions of pounds)

Plastic Resin	Transpor- tation	%	Packaging	%	Building and Construction	%	Electrical and Electronic
Thermosets							
Epoxy	–	–	–	–	–	–	66
Polyester	543	24.2	–	–	46	6.5	47
Urea and Melamine	–	–	62	0.3	–	–	–
Phenolics	54	2.4	–	–	91	12.8	160
Other Thermosets	74[a]	3.3	16	0.1	7	1.0	59
Total Thermosets	670	29.9	78	0.5	144	20.3	332
Thermoplastics							
LDPE	81[b]	3.6	7,312	45.1	72[b]	10.1	339
HDPE	–	–	3,519	21.7	–	–	111
Polypropylene	381	17.0	1,157	7.1	–	–	219
ABS and SAN	285	12.7	64	0.4	3	0.4	224
Polystyrene	–	–	2,025	12.5	41	5.8	278
Nylon	67	3.0	48	0.3	–	–	54
PVC	271	12.1	739	4.6	330	46.5	349
PVAc and Other Vinyls	–	–	–	–	–	–	–
Thermopolyester	–	–	771	4.8	–	–	–
Other Thermoplas- tics	74[a]	3.3	338	2.1	119[a]	16.8	372
Total Thermoplastics	1,159	51.7	15,973	98.6	565	79.6	1,947
Total Without Poly- urethane Foam	1,830	81.6	16,070	99.2	709	99.9	2,278
Polyurethane Foam	408[c]	18.2	137	0.8	1[c]	0.1	79
Total	2,242	100	16,206	100	710	100	2,357
Percent	6.4	–	46	–	2.0	–	6.7

[a]Data do not distinguish between other thermosets and other thermoplastics.
[b]Data do not distinguish between LDPE and HDPE.
[c]Data do not distinguish between polyurethane foam and other polyurethane uses.
Notes: Low Density Polyethylene (LDPE); High Density Polyethylene (HDPE); Acrylonitrile-Butadiene-Styrene (ABS); Syrene-Acrylonitrile (SAN); Polyvinyl Chloride (PVC); Polyvinyl Acetate (PVAc).

%	Furniture	%	Consumer and Institutional	%	Industrial Machinery	%	Adhesives and Other	%	Total	%
2.8	–	–	–	–	–	–	139	2.2	205	0.6
2.2	24	0.9	84	2.1	–	–	114	1.8	858	2.5
–	188	7.1	–	–	–	–	127	2.0	377	1.1
6.8	26	1.0	48	1.2	25	6.9	108	1.7	512	1.5
2.5	–	–	84	2.1	56	15.5	512	7.9	808	2.3
14.1	238	9.0	216	5.4	81	22.4	1,000	15.5	2,759	7.9
14.4	16	0.6	587	14.7	–	–	601	9.3	9,008	25.8
4.7	–	–	551	13.8	94	26.0	753	11.7	5,028	14.4
9.3	846	31.9	607	15.2	51	14.2	759	11.8	4,020	11.5
9.5	–	–	176	4.4	–	–	234	3.6	986	2.8
11.8	64	2.4	859	21.5	–	–	651	10.1	3,917	11.2
2.3	7	0.3	–	–	24	6.6	38	0.6	238	0.7
14.8	177	6.7	308	7.7	17	4.8	702	10.9	2,893	8.3
–	43	1.6	60	1.5	–	–	791	12.3	894	2.6
–	–	–	12	0.3	–	–	13	0.2	796	2.3
15.8	344	13.0	619	15.5	94	26.0	778	12.1	2,738	7.8
82.6	1,499	56.5	3,779	94.6	279	77.6	5,325	82.5	30,526	87.3
96.7	1,737	65.5	3,995	100	360	100	6,325	98.0	33,304	95.2
3.3	914	34.5	–	–	–	–	126	2.0	1,665	4.8
100	2,651	100	3,995	100	360	100	6,451	100	34,972	100
–	7.6	–	11.4	–	1.0	–	18.4	–	100	–

TABLE B.11. Projected Postconsumer Plastic Wastes: 1995 (millions of pounds)

Plastic Resin	Transpor-tation	%	Packaging	%	Building and Construction	%	Electrical and Electronic
Thermosets							
Epoxy	–	–	–	–	–	–	79
Polyester	326	14.4	–	–	258	6.7	57
Urea and Melamine	–	–	77	0.4	–	–	–
Phenolics	160	7.1	–	–	228	5.9	191
Other Thermosets	39	1.7	19	0.1	173	4.5	71
Total Thermosets	525	23.2	96	0.5	659	17.1	398
Thermoplastics							
LDPE	–	–	8,712	45.1	351[b]	9.1	406
HDPE	125	5.5	4,193	21.7	–	–	133
Polypropylene	302	13.3	1,379	7.1	–	–	261
ABS and SAN	275	12.1	77	0.4	195	5.0	267
Polystyrene	12	0.5	2,413	12.5	137	3.5	332
Nylon	111	4.9	57	0.3	–	–	65
PVC	154	6.8	881	4.6	1,900	49.1	419
PVAc and Other Vinyls	–	–	–	–	–	–	–
Thermopolyester	39	1.7	919	4.8	–	–	–
Other Thermoplastics	410	18.1	402	2.1	470[a]	12.2	444
Total Thermoplastics	1,659	62.9	19,052	98.6	3,053	78.9	2,327
Total Without Polyurethane Foam	1,951	86.1	19,147	99.1	3,712	96.0	2,722
Polyurethane Foam	315[c]	13.9	172	0.9	156[c]	4.0	76
Total	2,266	100	19,319.3	100	3,868	100	2,798
Percent	5.2	–	44.5	–	8.9	–	6.4

[a]Data do not distinguish between other thermosets and other thermoplastics.

[b]Data do not distinguish between LDPE and HDPE.

[c]Data do not distinguish between polyurethane foam and other polyurethane uses.

Notes: Low Density Polyethylene (LDPE); High Density Polyethylene (HDPE); Acrylonitrile-Butadiene-Styrene (ABS); Syrene-Acrylonitrile (SAN); Polyvinyl Chloride (PVC); Polyvinyl Acetate (PVAc).

%	Furniture	%	Consumer and Institutional	%	Industrial Machinery	%	Adhesives and Other	%	Total	%
2.8	–	–	–	–	–	–	156	2.2	235	0.5
2.0	2	0.1	92	2.1	–	–	127	1.8	862	2.0
–	–	–	–	–	–	–	142	2.0	219	0.5
6.8	–	–	53	1.2	30	6.9	120	1.7	782	1.8
2.5	178	5.7	92	2.1	67	15.5	574	7.9	1,213	2.8
14.2	180	5.8	237	5.4	97	22.4	1,119	15.5	3,311	7.6
14.5	–	–	645	14.7	–	–	673	9.3	10,787	24.8
4.8	–	–	606	13.8	113	26.0	843	11.7	6,013	13.8
9.3	981	31.2	667	15.2	61	14.2	850	11.8	4,501	10.4
9.5	6	0.2	193	4.4	–	–	262	3.6	1,275	2.9
11.9	46	1.5	944	21.5	–	–	729	10.1	4,613	10.6
2.3	8	0.3	–	–	29	6.6	42	0.6	312	0.7
15.0	288	9.2	338	7.7	21	4.8	786	10.9	4,787	11.0
–	60	1.9	66	1.5	–	–	885	12.3	1,011	2.3
–	–	–	13	0.3	–	–	14	0.2	985	2.3
15.9	431	13.7	680	15.5	113	26.0	871	12.1	3,821	8.8
83.2	1,820	57.8	4,152	94.6	336	77.6	5,962	82.5	38,361	88.3
97.3	2,003	63.6	4,389	100	433	100	7,081	98.0	41,438	95.4
2.7	1,144	36.4	–	–	–	–	142	2.0	2,005	4.6
100	3,147	100	4,389	100	433	100	7,223	100	43,443	100
–	7.2	–	10.1	–	1.0	–	16.6	–	100	–

Appendix C

Detailed Information on U.S. Automobile Shredder Operations

TABLE C.1. U.S. Shredder Operations (arranged by population)

State	City	Population	Company	Capacity	Installed	Total
Illinois	Chicago	3005072	Proler	300000	1963	
Illinois	Chicago	3005072	Il. Scrap Processing Co.	100000	1980	
Illinois	Chicago	3005072	General Iron Industries	80000	1980	
Illinois	Chicago	3005072	Pielet Bros. Iron & Metal	66000	1971	
Penn.	Philadelphia	1688210	Mayer Pollack	72000	1968	
Texas	Houston	1595138	Proler	300000	1965	
Texas	Houston	1595138	Newell Salvage	72000	1970	
Texas	Houston	1595138	Houston Junk Co.	24000	1974	
Michigan	Detroit	1203339	Auto Shred	72000	1975	
Michigan	Detroit	1203339	Mason Iron & Metal	24000	1969	
New York	Queens[a]	1000000	Proler	300000	1969	
New York	Bronx[a]	1000000	Bronx Industrial Scrap	72000	1979	
Texas	Dallas	904078	Comsteel, Inc.	48000	1970	
Michigan	Grand Rapids	811843	Berman Brothers	60000	1979	
Arizona	Phoenix	789704	National Metals Co.	48000	1969	
Maryland	Baltimore	786775	United Iron and Metal Co.	72000	1970	
Texas	San Antonio	785880	Newell Salvage	48000	1970	
Indiana	Indianapolis	700807	Auto Shred	85000	1973	
Tennessee	Memphis	646356	Mid-American Recycling	90000	1976	
Tennessee	Memphis	646356	Sanitized Steel Co.	54000	1974	
D.C.	Washington	638333	Joseph Smith & Sons	72000	1969	
Wisconsin	Milwaukee	636212	Miller Compressing	108000	1976	
Wisconsin	Milwaukee	636212	Afram Bros.	72000	1971	
California	San Jose	629442	Pan American Scrap	7500	1979	
Ohio	Cleveland	573822	Lurla Bros.	160000	1967	

(continued)

TABLE C.1. *Continued*

State	City	Population	Company	Capacity	Installed	Total
Ohio	Columbus	564871	Lurla Bros.	48000	1970	2946500
Mass.	Boston	562944	Proler	300000	1965	
Louisiana	New Orleans	557515	Southern Scrap Material	60000	1971	
Florida	Jacksonville	540920	Automotive Disposal	72000	1977	
Florida	Jacksonville	540920	David Joseph Co.	60000	1971	
Colorado	Denver	492365	Denver Metals	60000	1977	
Tennessee	Nashville	455651	Steiner-Liff	50000	1975	
Missouri	St. Louis	453085	L & M Shredding	48000	1979	
Missouri	Kansas City	448159	Kansas City Recycling	48000	1977	
Texas	El Paso	425259	Newell Salvage	30000	1971	
Georgia	Atlanta	425022	Newell Inds. of Atlanta	72000	1977	
Georgia	Atlanta	425022	London Iron & Metal Co.	60000	1971	
Georgia	Atlanta	425022	Central Iron & Steel	48000	1979	
Penn.	Pittsburgh	423938	Carco, Inc.	65000	1975	
Penn.	Pittsburgh	423938	Consolidated Iron	60000	1974	
Oklahoma	Oklahoma City	403313	Standard Iron & Metal	60000	1972	
Oklahoma	Oklahoma City	403313	Oklahoma Metal Processing	48000	1971	
Ohio	Cincinnati	385457	I. Deutsch	72000	1976	
Ohio	Cincinnati	385457	Cincinnati Auto Shred	72000	1976	
Oregon	Portland	366383	Schnitzer	72000	1970	
California	Long Beach	361334	Clean Steel, Inc.	180000	1967	
Oklahoma	Tulsa	360919	Tulsa Metal Processing	40000	1970	
New York	Buffalo	357870	Advance Metals	72000	1976	
Ohio	Toledo	354635	Kasle Bros.	54000	1967	

State	City		Company			
Texas	Austin	345469	Newell Salvage	30000	1971	
California	Oakland	339337	Schnitzer Steel	180000	1968	
N. Mexico	Albuquerque	331767	Albuquerque Metals	36000	1967	
Arizona	Tucson	330537	National Iron & Steel	20000	1966	
N. Jersey	Newark	329284	Nimco Shredding	72000	1973	
N. Carolina	Charlotte	314447	Southern Metals	54000	1971	
Kentucky	Louisville	298451	River City Shredding Co.	66000	1977	
Kentucky	Louisville	298451	Louisville Scrap Metal	36000	1967	
Alabama	Birmingham	284413	Shredders, Inc.	72000	1975	
Alabama	Birmingham	284413	National Tire and Salvage	72000	1972	
Alabama	Birmingham	284413	National Tire and Salvage	54000	1976	
Kansas	Wichita	279272	Kansas Metals Co.	48000	1975	
California	Sacramento	275741	Levin Metals	80000	1975	
Florida	Tampa	271523	David Joseph Co.	36000	1974	
Minnesota	St. Paul	270320	North Star Steel	120000	1973	
Minnesota	St. Paul	270320	H. S. Kaplan	66000	1969	2253000
Texas	Corpus Christi	231999	Commercial Metals	30000	1972	
Texas	Corpus Christi	231999	Industrial Salvage	30000	1975	
N. Jersey	Jersey City	223532	Prolerized Schiabo	300000	1968	
Louisiana	Baton Rouge	219419	Southern Scrap Material	60000	1974	
California	Anaheim	219311	Orange Co. Steel	65000	1981	
Virginia	Richmond	219214	Peck Iron & Metal Co.	72000	1974	
Louisiana	Shreveport	205820	Southwestern Iron Corp.	72000	1975	
Mississippi	Jackson	202895	Klean Steel	66000	1970	
Alabama	Mobile	200452	Pinto Island Metal Co.	48000	1970	
Tennessee	Knoxville	175030	Southern Foundry Supply	48000	1973	
Washington	Spokane	171300	American Recycling	48000	1968	
Wisconsin	Madison	170616	H. Samuels Co.	60000	1972	
New York	Syracuse	170105	Roth Steel Corp.	30000	1967	

(continued)

175

TABLE C.1. *Continued*

State	City	Population	Company	Capacity	Installed	Total
Tennessee	Chattanooga	169565	Southern Foundry Supply	48000	1970	
Nevada	Las Vegas	164674	Nevada Recycling	60000	1971	
Utah	Salt Lake C.	163033	Learner-Pepper	48000	1970	
Mass.	Worchester	161799	Steelmet	60000	1977	
Kansas	Kansas City	161087	Proler	300000	1968	
Kansas	Kansas City	161087	Kansas City Recycling	67000	1977	
Washington	Tacoma	158501	General Metals	72000	1966	
Washington	Tacoma	158501	Joseph Simon & Sons, Inc.	48000	1975	
Arkansas	Little Rock	158461	A. Tenenbaum	60000	1979	
Rhode Is.	Providence	156804	Metals Processing Co.	60000	1969	
California	Stockton	149779	Learner Co.[c]	80000	1980	
Georgia	Savannah	141390	Chatham Iron & Metal	36000	1974	
Connecticut	Hartford	136392	Suismam & Blumenthal	48000	1976	
Michigan	Lansing	130414	Summit Steel Proc. Corp.	48000	1974	
Michigan	Lansing	130414	Summit Steel Proc. Corp.	48000	1966	
Illinois	Peoria	124160	I. Bork & Son	120000	1975	
Illinois	Peoria	124160	A. Miller	50000	1965	
Illinois	Peoria	124160	Allied Iron & Steel Corp.	36000	1967	
Texas	Beaumont	118102	Southern Iron & Metal	48000	1971	
Virginia	Chesapeake	114486	Inter Coastal Steel	48000	1967	
Arizona	Tempe	106743	Marathon Steel	100000	1975	
California	Bakersfield	105611	Golden State Metals	72000	1966	
Iowa	Davenport	103264	Alter Company	72000	1977	
Virginia	Alexandria	103217	Alexandria Scrap Corp.	54000	1968	
New York	Albany	101727	Ute Industries	78000	1976	

State	ID	City	Company	Capacity	Year
				2744000	
Texas	101261	Waco	Lipsitz	54000	1975
Texas	98315	Abilene	Pine St. Salvage	24000	1970
Mass.	95172	Brockton	Brisco Baling	48000	1977
Ohio	94703	Canton	Luntz Corp.	120000	1968
Wisconsin	87899	Green Bay	Behen Processing Co.	60000	1977
N. Jersey	84910	Camden	Camden Iron & Metal	66000	1975
Michigan	79722	Kalamazoo	Superior Metals	85000	1978
Penn.	78686	Temple	Simon-Eastern Corp.	75000	1980
Michigan	77568	Taylor	Huron Valley	160000	1967
Michigan	77508	Saginaw	Ferro-Met	30000	1969
Michigan	76715	Pontiac	Fragment Products	120000	1979
Michigan	76715	Pontiac	Sam Allen & Son	60000	1968
Iowa	75985	Waterloo	Weissman Iron & Metal Co.	48000	1975
Arkansas	71626	Fort Smith	Yaffe Iron & Steel	72000	1976
S. Carolina	58242	Greenville	Industrial Scrap Inc.	48000	1968
Florida	57619	Pensacola	Auto Shred Ind.	72000	1975
Louisiana	57597	Monroe	Auto Shred of Louisiana	60000	1976
Iowa	56449	Council Bluff	Alter Company	72000	1976
California	54951	Redwood City	Levin Metals	88000	1976
California	54951	Redwood City	Learner Co.	72000	1971
Ohio	53927	Mansfield	Luntz	60000	1975
California	50000	Terminal Is.[b]	Proler	300000	1966
Illinois	50000	National City[b]	St. Louis Auto Shredding	150000	1975
California	50000	Etiwanda[b]	Ferromet, Inc.	96000	1971
New York	50000	N. Chili[b]	Union Processing	72000	1975
Mass.	50000	Wooster[b]	Stellmet	60000	1977
Michigan	50000	Sturgis[b]	Sturgis Iron	60000	1973
Michigan	50000	Iron Mount[b]	E. Kingsford Iron & Metal	48000	1974
Virginia	50000	Montvale[b]	Shredder Prod. Corp.	36000	1967
				2262000	

(continued)

TABLE C.1. *Continued*

State	City	Population	Company	Capacity	Installed	Total
Tennessee	Jackson	49131	H. O. Forgy & Son Inc.	72000	1975	
California	National City	48772	Scrap Disposal, Inc.	180000	1968	
Indiana	Kokomo	47808	Universal Steel, Inc.	72000	1975	
Georgia	Augusta	47532	Auto Recycling	66000	1975	
Georgia	Athens	42549	Loef Company	66000	1976	
Mississippi	Greenville	40613	Friedman Iron & Steel	48000	1971	
Michigan	Jackson	39739	Glick Iron & Metal	48000	1970	
Wisconsin	Fond Du Lac	35863	Sadoff & Rudoy	120000	1974	
Illinois	Alton	34171	Hyman-Michaels Co.	84000	1969	
Maryland	Hagerstown	34132	Conservit, Inc.	54000	1975	
Penn.	Williamsport	33401	Simon Wrecking	50000	1974	
Penn.	Williamsport	33401	Eastern Scrap Baling Co.	48000	1974	
Washington	Renton	30612	Sternoff	60000	1968	
Florida	Bradenton	30170	Manatee Metals	66000	1976	
Kentucky	Ashland	20764	Mansbach Metal Co.	70000	1980	
New York	Lindenhurst	26919	D & A Scrap	30000	1978	
Michigan	Holland	26281	Louis Padnos	48000	1972	1242000
S. Carolina	Lexington	25820	Owen Industries	60000	1974	
Rhode Is.	Johnson	24907	Metal Recycling Inc.	60000	1980	
Mass.	Tewksbury	24635	Tewksbury Auto Parts	48000	1968	
Mass.	Tewksbury	24635	Tewksbury Auto Parts	24000	1974	
S. Carolina	Darlington	23469	Darlington Metals	48000	1974	
New York	Lackawanna	22701	Robin Scrap Prod.	30000	1967	
Connecticut	North Haven	22080	M. Schiavone & Sons	66000	1968	
Kentucky	Newport	21586	Kirschner	66000	1976	
Texas	Eagle Pass	21407	Guadalupe Industrial	72000	1974	

178

State	City	Population	Company	Capacity	Year	
Texas	Eagle Pass	21407	Newell Salvage	36000	1974	1387000
Ohio	Wooster	19289	Wooster Iron & Metal	48000	1978	
Mass.	Greenfield	18436	I. Kramer	66000	1976	
Mass.	Greenfield	18436	Shredmetal Inc.	66000	1976	
N. Carolina	Statesville	16622	L. Gordon Iron & Metal	48000	1976	
Tennessee	Pulaski	15652	Denbo Scrap	80000	1971	
N. Carolina	Kernersville	15459	United Auto Disposal	60000	1970	
California	Castroville	14564	A & S Metals	15000	1980	
Florida	Opa Locka	14460	Scrap Metal Proc. Corp.	66000	1975	
Oregon	McMinnville	14080	Cascade Steel	72000	1975	
New York	Westbury	13871	Attonito Co.	72000	1976	
Illinois	Riverdale	13233	Fritz Enterprises	72000	1976	
Ohio	W. Carrolton	13148	Metal Shredders	50000	1971	
Penn.	Bever Falls	12525	Luria Bros.	60000	1977	
S. Carolina	Dillon	11957	Lockany Steel	48000	1974	
Florida	Cocoa Beach	10926	Daytona Salvage	48000	1979	
Tennessee	Rockwood	10023	Rockwood Iron & Metal Co.	66000	1972	
Texas	Midlothian	6154	Chaparral Steel	200000	1975	
Georgia	Union City	4780	Southern Foundry Supply	72000	1974	
Illinois	S. Beloit	3346	William Lans Sons Co.	54000	1970	
Michigan	Clare	3300	Mid-Michigan Recycle	60000	1975	
N. Hamp.	Madbury	987	Tewksbury Auto Parts	60000	1976	
Utah	Plymouth	238	David Joseph	120000	1981	566000
Totals		13400500				13400500

[a]Assumed part of New York City with population in excess of one million.
[b]Population not given in Andriot (1983); assumed to be 50,000.
[c]Capacity assumed by author. Capacity corresponds to other plants with similar horsepower.
Sources: Information on shredder operations: *Scrap Age* (1980) (arranged by author) (used with permission); population data: Andriot (1983).

TABLE C.2. U.S. Shredder Operations (arranged by capacity)

State	City	Population	Company	Capacity	Installed
Illinois	Chicago	3005072	Proler	300000	1963
Texas	Houston	1595138	Proler	300000	1965
Mass.	Boston	562944	Proler	300000	1965
California	Terminal Is.[a]	50000	Proler	300000	1966
Kansas	Kansas City	161087	Proler	300000	1968
N. Jersey	Jersey City	223532	Prolerized Schiabo	300000	1968
New York	Queens[b]	1000000	Proler	300000	1969
Texas	Midlothian	6154	Chaparral Steel	200000	1975
California	Long Beach	361334	Clean Steel, Inc.	180000	1967
California	National City	48772	Scrap Disposal, Inc.	180000	1968
California	Oakland	339337	Schnitzer Steel	180000	1968
Michigan	Taylor	77568	Huron Valley	160000	1967
Ohio	Cleveland	573822	Lurla Bros.	160000	1967
Illinois	National City[a]	50000	St. Louis Auto Shredding	150000	1975
Ohio	Canton	94703	Luntz Corp.	120000	1968
Minnesota	St. Paul	270320	North Star Steel	120000	1973
Wisconsin	Fond Du Lac	35863	Sadoff & Rudoy	120000	1974
Illinois	Peoria	124160	I. Bork & Son	120000	1975
Michigan	Pontiac	76715	Fragment Products	120000	1979
Utah	Plymouth	238	David Joseph	120000	1981
Wisconsin	Milwaukee	636212	Miller Compressing	108000	1976
Arizona	Tempe	106743	Marathon Steel	100000	1975
Illinois	Chicago	3005072	Il. Scrap Processing Co.	100000	1980
California	Etiwanda[a]	50000	Ferromet, Inc.	96000	1971
Tennessee	Memphis	646356	Mid-American Recycling	90000	1976

State	City		Company		Year
California	Redwood City	54951	Levin Metals	88000	1976
Indiana	Indianapolis	700807	Auto Shred	85000	1973
Michigan	Kalamazoo	79722	Superior Metals	85000	1978
Illinois	Alton	34171	Hyman-Michaels Co.	84000	1969
Tennessee	Pulaski	15652	Denbo Scrap	80000	1971
California	Sacramento	275741	Levin Metals	80000	1975
Illinois	Chicago	3005072	General Iron Industries	80000	1980
California	Stockton	149779	Learner Co.[c]	80000	1980
New York	Albany	101727	Ute Industries	78000	1976
Penn.	Temple	78686	Simon-Eastern Corp.	75000	1980
Washington	Tacoma	158501	General Metals	72000	1966
California	Bakersfield	105611	Golden State Metals	72000	1966
Penn.	Philadelphia	1688210	Mayer Pollack	72000	1968
D.C.	Washington	638333	Joseph Smith & Sons	72000	1969
Texas	Houston	1595138	Newell Salvage	72000	1970
Oregon	Portland	366383	Schnitzer	72000	1970
Maryland	Baltimore	786775	United Iron and Metal Co.	72000	1970
California	Redwood City	54951	Learner Co.	72000	1971
Wisconsin	Milwaukee	636212	Afram Bros.	72000	1971
Alabama	Birmingham	284413	National Tire and Salvage	72000	1972
N. Jersey	Newark	329284	Nimco Shredding	72000	1973
Virginia	Richmond	219214	Peck Iron & Metal Co.	72000	1974
Texas	Eagle Pass	21407	Guadalupe Industrial	72000	1974
Georgia	Union City	4780	Southern Foundry Supply	72000	1974
Oregon	McMinnville	14080	Cascade Steel	72000	1975
Alabama	Birmingham	284413	Shredders, Inc.	72000	1975
New York	N. Chili[a]	50000	Union Processing	72000	1975
Michigan	Detroit	1203339	Auto Shred	72000	1975
Tennessee	Jackson	49131	H. O. Forgy & Son. Inc.	72000	1975

(continued)

TABLE C.2. *Continued*

State	City	Population	Company	Capacity	Installed
Indiana	Kokomo	47808	Universal Steel, Inc.	72000	1975
Florida	Pensacola	57619	Auto Shred Ind.	72000	1975
Louisiana	Shreveport	205820	Southwestern Iron Corp.	72000	1975
Iowa	Council Bluff	56449	Alter Company	72000	1976
Arkansas	Fort Smith	71626	Yaffe Iron & Steel	72000	1976
New York	Westbury	13871	Attonito Co.	72000	1976
New York	Buffalo	357870	Advance Metals	72000	1976
Ohio	Cincinnati	385457	Cincinnati Auto Shred	72000	1976
Illinois	Riverdale	13233	Fritz Enterprises	72000	1976
Ohio	Cincinnati	385457	I. Deutsch	72000	1976
Florida	Jacksonville	540920	Automotive Disposal	72000	1977
Iowa	Davenport	103264	Alter Company	72000	1977
Georgia	Atlanta	425022	Newell Inds. of Atlanta	72000	1977
New York	Bronx[b]	1000000	Bronx Industrial Scrap	72000	1979
Kentucky	Ashland	27064	Mansbach Metal Co.	70000	1980
Kansas	Kansas City	161087	Kansas City Recycling	67000	1977
Connecticut	North Haven	22080	M. Schiavone & Sons	66000	1968
Minnesota	St. Paul	270320	H. S. Kaplan	66000	1969
Mississippi	Jackson	202895	Klean Steel	66000	1970
Illinois	Chicago	3005072	Pielet Bros. Iron & Metal	66000	1971
Tennessee	Rockwood	10023	Rockwood Iron & Metal Co.	66000	1972
N. Jersey	Camden	84910	Camden Iron & Metal	66000	1975
Florida	Opa Locka	14460	Scrap Metal Proc. Corp.	66000	1975
Georgia	Augusta	47532	Auto Recycling	66000	1975
Kentucky	Newport	21586	Kirschner	66000	1976
Mass.	Greenfield	18436	I. Kramer	66000	1976

182

State	City		Company		Year
Florida	Bradenton	30170	Manatee Metals	66000	1976
Mass.	Greenfield	18436	Shredmetal Inc.	66000	1976
Georgia	Athens	42549	Loef Company	66000	1976
Kentucky	Louisville	298451	River City Shredding Co.	66000	1977
Penn.	Pittsburgh	423938	Carco, Inc.	65000	1975
California	Anaheim	219311	Orange Co. Steel	65000	1981
Michigan	Pontiac	76715	Sam Allen & Son	60000	1968
Washington	Renton	30612	Sternoff	60000	1968
Rhode Is.	Providence	156804	Metals Processing Co.	60000	1969
N. Carolina	Kernersville	15459	United Auto Disposal	60000	1970
Louisiana	New Orleans	557515	Southern Scrap Material	60000	1971
Florida	Jacksonville	540920	David Joseph Co.	60000	1971
Nevada	Las Vegas	164674	Nevada Recycling	60000	1971
Georgia	Atlanta	425022	London Iron & Metal Co.	60000	1971
Wisconsin	Madison	170616	H. Samuels Co.	60000	1972
Oklahoma	Oklahoma City	403313	Standard Iron & Metal	60000	1972
Michigan	Sturgis[a]	50000	Sturgis Iron	60000	1973
Louisiana	Baton Rouge	219419	Southern Scrap Material	60000	1974
Penn.	Pittsburgh	423938	Consolidated Iron	60000	1974
S. Carolina	Lexington	25820	Owen Industries	60000	1974
Ohio	Mansfield	53927	Luntz	60000	1975
Michigan	Clare	3300	Mid-Michigan Recycle	60000	1975
Louisiana	Monroe	57597	Auto Shred of Louisiana	60000	1976
N. Hamp.	Madbury	987	Tewksbury Auto Parts	60000	1976
Wisconsin	Green Bay	87899	Behen Processing Co.	60000	1977
Mass.	Wooster[a]	50000	Steelmet	60000	1977
Penn.	Bever Falls	12525	Luria Bros.	60000	1977
Colorado	Denver	492365	Denver Metals	60000	1977
Mass.	Worchester	161799	Steelmet	60000	1977

(continued)

183

TABLE C.2. *Continued*

State	City	Population	Company	Capacity	Installed
Michigan	Grand Rapids	811843	Berman Brothers	60000	1979
Arkansas	Little Rock	158461	A. Tenenbaum	60000	1979
Rhode Is.	Johnson	24907	Metal Recycling Inc.	60000	1980
Ohio	Toledo	354635	Kasle Bros.	54000	1967
Virginia	Alexandria	103217	Alexandria Scrap Corp.	54000	1968
Illinois	S. Beloit	3346	William Lans Sons Co.	54000	1970
N. Carolina	Charlotte	314447	Southern Metals	54000	1971
Tennessee	Memphis	646356	Sanitized Steel Co.	54000	1974
Maryland	Hagerstown	34132	Conservit, Inc.	54000	1975
Texas	Waco	101261	Lipsitz	54000	1975
Alabama	Birmingham	284413	National Tire and Salvage	54000	1976
Illinois	Peoria	124160	A. Miller	50000	1965
Ohio	W. Carrolton	13148	Metal Shredders	50000	1971
Penn.	Williamsport	33401	Simon Wrecking	50000	1974
Tennessee	Nashville	455651	Steiner-Liff	50000	1975
Michigan	Lansing	130414	Summit Steel Proc. Corp.	48000	1966
Virginia	Chesapeake	114486	Inter Coastal Steel	48000	1967
S. Carolina	Greenville	58242	Industrial Scrap Inc.	48000	1968
Washington	Spokane	171300	American Recycling	48000	1968
Mass.	Tewksbury	24635	Tewksbury Auto Parts	48000	1968
Arizona	Phoenix	789704	National Metals Co.	48000	1969
Alabama	Mobile	200452	Pinto Island Metal Co.	48000	1970
Utah	Salt Lake C.	163033	Learner-Pepper	48000	1970
Texas	Dallas	904078	Comsteel, Inc.	48000	1970
Michigan	Jackson	39739	Glick Iron & Metal	48000	1970
Tennessee	Chattanooga	169565	Southern Foundry Supply	48000	1970

184

Texas	San Antonio	785880	Newell Salvage	48000	1970
Ohio	Columbus	564871	Lurla Bros.	48000	1970
Texas	Beaumont	118102	Southern Iron & Metal	48000	1971
Oklahoma	Oklahoma City	403313	Oklahoma Metal Processing	48000	1971
Mississippi	Greenville	40613	Friedman Iron & Steel	48000	1971
Michigan	Holland	26281	Louis Padnos	48000	1972
Tennessee	Knoxville	175030	Southern Foundry Supply	48000	1973
S. Carolina	Dillon	11957	Lockany Steel	48000	1974
Michigan	Lansing	130414	Summit Steel Proc. Corp.	48000	1974
S. Carolina	Darlington	23469	Darlington Metals	48000	1974
Penn.	Williamsport	33401	Eastern Scrap Baling Co.	48000	1974
Michigan	Iron Mount[a]	50000	E. Kingsford Iron & Metal	48000	1974
Iowa	Waterloo	75985	Weissman Iron & Metal Co.	48000	1975
Washington	Tacoma	158501	Joseph Simon & Sons, Inc.	48000	1975
Kansas	Wichita	279272	Kansas Metals Co.	48000	1975
N. Carolina	Statesville	16622	L. Gordon Iron & Metal	48000	1976
Connecticut	Hartford	136392	Suismam & Blumenthal	48000	1976
Mass.	Brockton	95172	Brisco Baling	48000	1977
Missouri	Kansas City	448159	Kansas City Recycling	48000	1977
Ohio	Wooster	19289	Wooster Iron & Metal	48000	1978
Georgia	Atlanta	425022	Central Iron & Steel	48000	1979
Missouri	St. Louis	453085	L & M Shredding	48000	1979
Florida	Cocoa Beach	10926	Daytona Salvage	48000	1979
Oklahoma	Tulsa	360919	Tulsa Metal Processing	40000	1970
Illinois	Peoria	124160	Allied Iron & Steel Corp.	36000	1967
N. Mexico	Albuquerque	331767	Albuquerque Metals	36000	1967
Virginia	Montvale[a]	50000	Shredder Prod. Corp.	36000	1967
Kentucky	Louisville	298451	Louisville Scrap Metal	36000	1967
Texas	Eagle Pass	21407	Newell Salvage	36000	1974

(continued)

TABLE C.2. *Continued*

State	City	Population	Company	Capacity	Installed
Florida	Tampa	271523	David Joseph Co.	36000	1974
Georgia	Savannah	141390	Chatham Iron & Metal	36000	1974
New York	Syracuse	170105	Roth Steel Corp.	30000	1967
New York	Lackawanna	22701	Robin Scrap Prod.	30000	1967
Michigan	Saginaw	77508	Ferro-Met	30000	1969
Texas	El Paso	425259	Newell Salvage	30000	1971
Texas	Austin	345469	Newell Salvage	30000	1971
Texas	Corpus Christi	231999	Commercial Metals	30000	1972
Texas	Corpus Christi	231999	Industrial Salvage	30000	1975
New York	Lindenhurst	26919	D & A Scrap	30000	1978
Michigan	Detroit	1203339	Mason Iron & Metal	24000	1969
Texas	Abilene	98315	Pine St. Salvage	24000	1970
Mass.	Tewksbury	24635	Tewksbury Auto Parts	24000	1974
Texas	Houston	1595138	Houston Junk Co.	24000	1974
Arizona	Tucson	330537	National Iron & Steel	20000	1966
California	Castroville	14564	A & S Metals	15000	1980
California	San Jose	629442	Pan American Scrap	7500	1979
Total				13400500	

[a]Population not given in Andriot (1983); assumed to be 50,000.

[b]Assumed part of New York City with population in excess of one million.

[c]Capacity assumed by author. Capacity corresponds to other plants with similar horsepower.

Sources: Information on shredder operations: *Scrap Age* (1980) (arranged by author) (used with permission); population data: Andriot (1983).

186

Bibliography

Abert, James G., Harvey Alter, and J. Frank Bernheisel. 1974. "The economics of resource recovery from municipal solid waste." *Science*, 183 (March), pp. 1052–58.

Adams, Robert L. 1972. "An economic analysis of the junk auto with emphasis on processing costs," in *Proceedings of the Third Mineral Waste Utilization Symposium*, sponsored by the U.S. Bureau of Mines and IIT Research Institute, held in Chicago, Illinois, March 14–16.

Aiken, W. E. 1976. "Outlook: Plastics recycling develops slowly, with separation methods the key to the future." *Purchasing*, June 8, p. 112.

Alvarez, Ronald J. 1985. "A look at U.S. plants that are burning MSW." *Waste Age*, January, pp. 58–60.

American Metal Market. 1979. "RSR to recover polypropylene from scrapped auto batteries." May 31, p. 7.

American Metal Market. 1984a. "U.S. cars to use 2.2 billion lbs of plastics by '95: Predicasts." July 30, p. 7.

American Metal Market. 1984b. "High-strength steel, plastic content will rise in autos." October 15, p. 21.

American Petroleum Institute. 1984. *Basic Petroleum Data Book* 4(1). Washington, D.C., January.

Analy-Syn Labs, Incorporated. 1977. "Plastics in wastes: Their potential value: Final report." DOE/SF/903 86-T1, prepared for the U.S. Department of Energy, Washington, D.C.

Anderson, Earl V. 1975. "Industry steps up efforts to recycle plastics wastes." *Chemical and Engineering News*, September 22, pp. 16–17.

Anderson, Robert C. 1977. "Public policies toward the use of scrap materials." *American Economic Review*, 67(1), pp. 355–58.

Andres, B. et al. 1982. "Recovery of raw materials by pyrolysis." BMFT-FB-T 81-017, October (translated into English), Bundesminister für Forschung und Technologie, Bonn, Germany.

Andriot, Johyn L., ed. 1983. *Population Abstracts of the United States*. McLean, Virginia: Andriot Associates.

Arthur D. Little, Incorporated. 1973. "Incentives for recycling and reuse of plastics: A summary report," prepared for the U.S. Environmental Protection Agency, Washington, D.C.

Automotive Engineering. 1979. "Automobile recycling offers renewable but changing resource." May, pp. 56–58.

Baller, J. 1982. "Japan's national waste recovery plan." *Waste Age*, August, pp. 65–69.

Barna, Bruce A., David R. Johnsrud, and Richard L. Lamparter. 1980. "Petrochemicals from waste: Recycling PET bottles." *Chemical Engineering*, December 1, pp. 50–51.

Basta, Nicholas. 1985. "A renaissance in recycling." *High Technology*, October, pp. 32–39.

Basta, Nicholas, et al. 1984. "Plastics recycle: A revival." *Chemical Engineering*, June 25, pp. 22–26.

Bauer, Stephen H. 1976. "Recycling thermoset materials," paper presented at the 34th Annual Technical Conference of the Society of Plastics Engineers, April 26–29, Atlantic City, N.J.

_____. 1977. "Processing plastics—Recycle thermoset scrap effectively." *Plastics Engineering*, March, pp. 44–46.

Baum, Bernard and Charles H. Parker. 1974. "Plastics waste management." October, DeBell and Richardson, Incorporated. Prepared for the Manufacturing Chemists Association, Washington, D.C.

Baumol, William J. 1977. "On recycling as a moot environmental issue." *J. Environ. Econ. Mgmt*, 4(1), pp. 83–87.

Bell, J. P., S. J. Huang, and J. R. Knox. 1974. "Synthesis and testing of polymers susceptible to hydrolysis by proteolytic enzymes." TR75-48-CE + MEL, Institute of Materials Science, University of Connecticut, prepared for the U.S. Army Natick Laboratories, Natick, MA.

Bennett, R. A. 1985. "PET recycling: Today's plastic soda bottle fills tomorrow's ski jacket." *Waste Age*, January, pp. 50–53.

Berry, Bryan H. 1983. "50 percent of today's car is steel. Tomorrow?" *Iron Age,* September 5, pp. 83–91.

Bever, Michael B. 1977. "Recycling in the materials system." *Materials and Society*, 1(2), pp. 167–76.

_____. 1980. "The impact of materials and design changes on the recycling of automobiles." *Materials and Society*, 4(3), pp. 375–85.

Bevis, M., N. Irving, and P. Allan. 1983. "Recovery and re-use of polymeric cable scrap." *Conservation and Recycling*, 6(½), pp. 3–10.

Bhatia, Jeet and Robert A. Rossi. 1982. "Pyrolysis process converts waste polymers to fuel oils." *Chemical Engineering*, October, pp. 58–59.

Bidwell, Robin and Karen Raymond. 1978. "Resource recovery in Europe." *NCRR Bulletin*, 8(1, Winter), pp. 3–9.

Bilbrey, J. H., Jr., J. W. Sterner, and E. G. Valdez. 1978. "Resource recovery from automobile shredder residues." Bureau of Mines, U.S. Department of the Interior, paper presented at the First World Recycling Congress, Swiss Industries Fair, March 6–7, Basel, Switzerland.

Bollard, A. E. 1982. "Plastics recycling and employment generation." *Plastics and Rubber International*, December, pp. 227–28.

Bradley, William W. and James J. Florio. 1980. "Financing resource recovery." *Waste Age*, July, pp. 38–39.

Brady, Jack D. and John H. Andros. 1983. "Gaseous and particulate emission control on industrial solid waste incinerators using wet scrubbers," paper presented at the 76th Annual Meeting of the Air Pollution Control Association, June 19–24, Atlanta, GA.

Bridgwater, A. J. and K. Lidgren, eds. 1981. *Household Waste Management in Europe: Economics and Techniques*. New York: Reinhold.

Broadman, Harry G. 1982. "The social cost of imported oil: Theoretical issues and empirical estimates," in *Energy Modeling IV: Planning for Energy Disruptions*. pp. 529-66. Chicago, IL: Institute of Gas Technology.

Brown, David G. 1979. "Recycling equipment for PET beverage bottles." *Plastics Machinery and Equipment*, October, pp. 32-33.

Buekens, A. G. 1977. "Some observations on the recycling of plastics and rubber." *Conservation and Recycling*, 1, pp. 247-71.

Burdick, T. E. and B. D. Bauman. 1984. "Ground scrap as a reactive foam extender." *Modern Plastics*, October, pp. 76-80.

Burlace, C. J. and L. Whalley. 1977. "Waste plastics and their potential for recycle." Warren Spring Laboratory, Stevenage, England.

Caputo, Dennis L. 1982. "Municipal plastic waste: A largely untapped resource." *Plastics Engineering*, February, pp. 33-35.

Chemical and Engineering News. 1981. "Reclaiming polyester bottles a big business." January 5, p. 30.

Chemical and Engineering News. 1985. "Rutgers gets plastic recycling institute." March 25, p. 31.

Chemical Engineering. 1982. "PET bottles get a tryout as boiler fuel." November 4, p. 95.

Chemical Engineering. 1984. "Plastics recycling: A revival." June 25, pp. 22-26.

Chemical Marketing Reporter. 1985. "Plastics recycling gets off the ground in Jersey: Major firms involved." March 25, p. 7.

Chemical Week. 1980. "Big plastics market in autos after 1985." February 20, p. 54.

Chemical Week. 1982a. "A plastic-body car for the mass market." August 4, p. 26.

Chemical Week. 1982b. "Plastic waste: A source for chemicals and fuels." August 11, p. 34.

Chemical Week. 1982c. "Plastics' robust hopes for the auto industry." September 1, pp. 12-14.

Chemical Week. 1984. "Reinforced plastics are a hit." January 25, pp. 11-12.

Chemtech. 1978. "New thermoplastics from old." August, pp. 502-08.

Coyle, Bernard H., Jr., Judith Koperski, and Robert N. Anderson. 1976. "A technical and economic analysis of processes for the recovery of metals in the nonferrous portion of automobile shredder refuse," in *Proceedings of the Fifth Mineral Waste Utilization Symposium*, sponsored by the U.S. Bureau of Mines and IIT Research Institute, held in Chicago, IL, April 13-14.

Criner, E. A. 1977. "An economic analysis of a plastic recycling program," paper presented at the American Institute of Industrial Engineers Spring Annual Conference, May 24-27, Dallas, TX.

Curlee, T. Randall. 1984a. "The economic feasibility of recycling plastic wastes: Preliminary assessment." ORNL/TM-9098, April, Oak Ridge, TN: Oak Ridge National Laboratory.

_____. 1984b. "Recycling plastic wastes: Quantity projections and preliminary competitive assessment." *Materials and Society*, 8(3), pp. 529-49.

_____. 1985a. "The recycle of plastics from auto shredder residue." *Materials and Society*, 9(1), pp. 29-43.

_____. 1985b. "Plastic wastes and the market penetration of auto shredders." *Tech-*

nological Forecasting and Social Change, 28(1), pp. 29–42.

_____. 1985c. "Background information for the short-term assessment of conservation RD projects," draft ORNL report, Oak Ridge National Laboratory, Oak Ridge, TN.

Curry, David T. 1980. "Plastics team up with metals for rigid, lightweight components." *Machine Design*, July, pp. 68–72.

Dean, K. C. and C. J. Chindgren. 1972. "Advances in technology for recycling obsolete cars," in *Proceeding of the Third Mineral Waste Utilization Symposium*, sponsored by the U.S. Bureau of Mines and IIT Research Institute, held in Chicago, IL, March 14–16.

Dean, K. C., J. W. Sterner, and E. G. Valdez. 1974. "Effect of increasing plastics content on recycling of automobiles." Technical Progress Report 79, May, Bureau of Mines Solid Waste Program, U.S. Department of the Interior, Washington, D.C.

Deanin, Rudolph D. and Chaitanya S. Nadkarni. 1984. "Recycling of the mixed plastics fraction from junked autos. I. Low-pressure molding." *Advances in Polymer Technology*, 4(2), pp. 173–76.

Diaz, Luis F., George M. Savage, and Clarence G. Golueke. 1982. *Resource Recovery from Municipal Solid Waste*. Volumes I and II. Boca Raton, FL: CRC Press.

Donovan, Richard C., Anthony J. Pompeo, and Emanuele Scalco. 1977. "Recycling PVC." *Bell Laboratory Record*, September, pp. 215–18.

Drain, K. F., W. R. Murphy, and M. S. Otterburn. 1981. *Conservation and Recycling*, 4(4), pp. 201–18.

Dreissen, H. H. and A. T. Basten. 1976. "Reclaiming products from shredding junked cars by the water-only and heavy-medium cyclone processes," in *Proceedings of the Fifth Mineral Waste Utilization Symposium*, sponsored by the U.S. Bureau of Mines and IIT Research Institute, held in Chicago, IL, April 13–14.

Ducey, R. A., D. M. Joncich, K. L. Griggs, and S. R. Sias. 1985. "20 Common problems found in small waste-to-energy plants." *Waste Age*, May, pp. 50–54.

Dunphy, Joseph F. 1985. "Pushing for increased plastics recycling." *Chemical Week*, March 27, pp. 16–17.

Energy User News. 1978. "Waste plastics converted to boiler fuel." November 20, p. 12.

Environmental Science and Technology. 1975. "Disposing of solid waste by pyrolysis." February, pp. 98–99.

Environmental Science and Technology. 1976. "It may now be possible to separate plastics from municipal wastes according to the type of plastic." September, p. 852.

European Plastics News. 1974. "Japanese process makes recycling profitable." November, pp. 32–34.

European Plastics News. 1977. "Recycling developments in France and Britain." August, p. 12.

Ewert, Norman James. 1976. "An economic analysis of incentives in automobile recycling." Ph.D. dissertation, July, Department of Economics, Southern Illinois Univ., Carbondale, IL.

Florida State Department of Environmental Regulation. 1983. "National directory of manufacturers utilizing recycled materials." NBS-GCR-83-424, report for the

U.S. National Bureau of Standards, Washington, D.C., March.

Fowler, James E. 1984. "Processing automobile hulks." *Resource Recycling*, 3(4), pp. 14–18.

Galiguzova, T. V. and M. N. Dmitrieva. 1983. "Refinement of secondary plastics processing technology." *Chemical Petroleum Engineering*, 19(7/8), pp. 314–15.

Ginter, Peter M. and Jack M. Starling. 1978. "Reverse distribution channels for recycling." *California Management Review*, 3, pp. 72–82.

Goddard, Haynes C. 1975. *Managing Solid Waste: Economics, Technology, and Institutions*. New York: Praeger Publishers.

Goodyear Tire and Rubber Company. Undated. "Recycling polyester bottles." *Cleartuf Facts*, CT-17, Akron, OH.

Grace, Richard, R. Kerry Turner, and Ingo Walter. 1978. "Secondary materials and international trade." *J. Environ. Econ. Mgmt.*, 5(2), pp. 172–86.

Gray, Ralph. 1972. "The economics of disposal pollution and recycling." *Q. Rev. Econ. Bus.*, 12(1), pp. 43–51.

Grubbs, M. R. and K. H. Ivey. 1972. "Recovering plastics from urban refuse by electrodynamic techniques." BM-/TP/R-63, December, U.S. Bureau of Mines, U.S. Department of the Interior, Washington D.C.

Halgren, Michael D. 1980. "Recycling and resource recovery: State and municipal legal impediments." *Columbia J. Environ. Law*, 7(Fall), pp. 1–31.

Hancock, Harvey G. and Philip Hubbauer. 1975. "Recycling turns scrap phones into new plastic products." *Bell Laboratory Record*, December, pp. 427–29.

Harwood, Julius J. 1977a. "Recycling the junk car — A case study of the automobile as a renewable resource." *Materials and Society*, 1(2), pp. 177–81.

Harwood, Julius J. 1977b. "Recycling the junk car." *Technology Review*, 79(4), pp. 32–37.

Hawkins, W. Lincoln. 1980. "Maximizing the life cycle of plastics: Final report." Plastics Institute of America, Hoboken, NJ, prepared for the U.S. Department of Energy, Washington, D.C.

Hawkins, W. Lincoln. 1982. "Progress report: Recycling of plastic scrap." Plastics Institute of America, prepared for the Energy Conservation and Utilization Technology Program, U.S. Department of Energy, Washington, D.C.

Henstock, Michael E. 1980. "Some barriers to the use of materials recovered from municipal solid waste." *Resources Policy*, 6(3), pp. 240–52.

Hoel, Michael. 1978. "Resource extraction and recycling with environmental costs." *J. Environ. Econ. Mgmt.*, 5(3), pp. 220–35.

Holcomb, M. C. and S. Koshy. 1984. "Transportation energy data book: Edition 7." ORNL-6050, June, Oak Ridge National Laboratory, Oak Ridge, TN, prepared for the Office of Vehicle and Engine Research and Development, U.S. Department of Energy, Washington, D.C.

Holman, J. L., J. B. Stephenson, and M. J. Adam. 1974. "Recycling of plastics from urban and industrial refuse." RI 7955, U.S. Department of the Interior, Washington, D.C.

Huang, C. J. and Charles Dalton. 1975. "Energy recovery from solid waste." NASA CR-2526, April, University of Houston, prepared for the U.S. National Aeronautics and Space Administration, Washington, D.C.

Huffman, George L. and Daniel J. Keller. 1973. "The plastics issue," in *Polymer Science and Technology*, edited by James Guillet, pp. 155–67. New York: Plenum Press.

Huls, Jon and Tom Archer. 1981. "Resource recovery from plastic and glass wastes." EPA-600/2-81-123, Pacific Environmental Services, Incorporated, prepared for the U.S. Environmental Protection Agency, Washington, D.C.

Hydrocarbon Processing. 1975. "Mitsubishi petrochemical is in plastic recycling business." August, p. 11.

Ilgenfritz, E. M. 1975. "Plastics waste handling practices in solid waste management." *Water, Air, and Soil Pollution*, 4(2), pp. 191–99.

Industronics, Incorporated. 1982. "Test burn report: PET bottles: 'Consertherm' Waste-to-Energy System." November 29, South Winsor, CT.

Industry Media, Incorporated. 1978. "Plastics: Villain or benefactor in our environment." Denver, CO.

Ingle, George W. 1973. "An industry view of plastics in the environment," in *Polymer Science and Technology*, edited by James Guillet, pp. 139–53. New York: Plenum Press.

Institute of Scrap Iron and Steel. Undated. "Metallic scrap: The manufactured resource." Washington, D.C.

International Research and Technology Corporation. 1973. "Recycling plastics: A survey and assessment of research and technology." June, prepared for the Society of the Plastics Industry, New York, NY.

Jensen, W. James, James L. Holman, and James B. Stephenson. 1974. "Recycling and disposal of waste plastics," in *Recycling and Disposal of Solid Waste: Industrial, Agricultural, Domestic*, edited by T. F. Yen, pp. 219–26. Ann Arbor, MI: Ann Arbor Science Publishers.

Johnson, Charles. 1985. "NSWMA's 1984 tipping fee survey." *Waste Age*, March, pp. 45–48.

Jones, Jerry, 1978. "Converting solid wastes and residues to fuel." *Chemical Engineering*, January 2, pp. 87–94.

Journal of Commerce. 1980. "Goodyear considers possible recycling and ecology of PET soda bottle." May 14, p. 5.

Journal of Commerce. 1981. "U.S. increases efforts to use scrap plastics." May 14, p. 10.

JRB Associates. 1981. "Solid waste data: A compilation of statistics on solid waste management within the United States" (no report no. given). Prepared for the Environmental Protection Agency, Washington, D.C.

Kaiser, E. R. and A. A. Carotti. 1971a. "Incineration of municipal refuse with 2% and 4% additions of four plastics." June 30, Chemical Engineering Department, New York Univ., New York, NY.

———. 1971b. "Incineration of municipal refuse with 2% and 4% additions of polyethylene terephthalate." December 17, Chemical Engineering Department, New York Univ., New York, NY.

Kaiser, R., R. P. Wasson, and A. C. W. Daniels. 1977. "Automobile scrappage and recycling industry study: Overview report." September, prepared for Transportation Systems Center, Cambridge, MA. H. H. Aerospace Design Co., Bedford, MA.

Kaminsky, W., J. Menzel, and H. Sinn. 1976. "Recycling of plastics." *Conservation and Recycling*, 1, pp. 91–110.

Lanza, Donna R. 1982. "Municipal solid waste regulation: An ineffective solution to a national problem." *Fordham Urban Law Journal*, 10(Spring), pp. 215–45.

Laswell, Dixie L. 1984. "State-federal relations under subtitle C of the resource conservation and recovery act." *Natural Resources Law*, 16(Winter), pp. 641–64.

Layard, P. R. G. and A. A. Walters. 1978. *Microeconomic Theory*. New York: McGraw Hill.

Lee, Samuel. 1979. "Waste plastic converted to fuel oil." *Chemical Processing*, April, pp. 40–42.

Leidner, Jacob. 1978. "Recovery of the value from postconsumer plastic waste." *Polymer-Plastics Technology and Engineering*, 10(2), pp. 199–205.

_____. 1981. *Plastics Waste: Recovery of Economic Value*. New York: Marcel Dekker.

Lock, Jim. 1978. "From waste polymers to durable board." *Processing*, 24(2), pp. 49–51.

Machacek, Raymond. 1983. "Economic and energy audit of converting waste plastics to oil." PNL-4734, Arthur D. Little, Inc., June. Prepared for the U.S. Department of Energy, Washington, D.C.

Machine Design. 1984. "Shredded automotive plastics promise improved particle board." February 23, p. 4.

Mahoney, L. R., J. Braslaw, and J. J. Harwood. 1979. "Automobile recycling offers renewable but changing resources." *Automotive Engineering*, 87(5), pp. 56–58.

Marynowski, Chester W. 1972. "Disposal of polymer solid wastes by primary polymer producers and plastics fabricators." EPA-SW-34C-72, Stanford Research Institute, prepared for the U.S. Environmental Protection Agency, Washington, D.C.: U.S. Government Printing Office.

Materials Engineering. 1978. "Recycle your junk and make usable plastic." July, p. 18.

McClellan, T. R. 1983. "Composite board from auto scrap (using polyisocyanates)." *Modern Plastics*, February, pp. 50–52.

Metal Scrap Research and Educational Foundation. 1983. "Research report: Shredder residue." December, Washington, D.C.

Midwest Research Institute. 1974. "Resource and environmental profile analysis of plastics and competitive materials." November, prepared for The Society of the Plastics Industry, New York, NY.

Milgrom, Jack. 1972. "Incentives for recycling and resuse of plastics." EPA-SW-41C-72, Arthur D. Little, Inc., prepared for the U.S. Environmental Protection Agency, Washington, D.C.

_____. 1973. "Identifying the nuisance plastics." *New Scientist*, January 25, pp. 184–86.

_____. 1979. "Recycling plastics: Current status." *Conservation and Recycling*, 3(3/4), pp. 327–35.

_____. 1982. "An overview of plastics recycling." *Polymer-Plastics Technology and Engineering*, 18(2), pp. 167–78.

_____. 1984. "Recycling the plastic beverage bottle." *Beverage World*, February, pp. 57–60.

_____. 1985. "Markets for recovered PET scrap." *Beverage World*, June, pp, 86–87.

Moch, John A. 1978. "Recycled materials: Looking for new uses." *Materials Engineering*, February, pp. 8–12.

Modern Plastics. 1977. "Industry and market news: Exploring the practicalities of recycling plastics." January, p. 110.

Modern Plastics. 1980a. "PET bottle recycling gets more practical." April, pp. 82–83.

Modern Plastics. 1980b. "Total recycling yields clean PET." September, p. 14.

Modern Plastics. 1983. "Low-cost materials from scrap: How they do it in Europe." May, pp. 56–57.

Modern Plastics International. 1984. "PET recycling plans can ease waste-disposal worries." May, p. 12.

Monsanto Research Corporation. 1981. "Waste Plastic Recycling Information Exchange: Volume I. Summary Report." NATO/CCMS 123-Volume I, Dayton, OH. Prepared for the NATO Commission on the challenges of Modern Society, Brussels, Belgium.

Morel, E., G. Richert, and C. Martin. 1980. "Examination of the possibilities of reusing thermosetting wastes of industrial origin." EUR-6833-EN, Commission of the European Communities, Luxembourg.

Morris, Cecil E., Jr. 1981. "Conserving natural resources: Toward a comprehensive state solid waste recycling program under the Federal Resource Conservation and Recovery Act." *New York Univ. Review of Law and Social Change*, 10(Summer); pp. 469–501.

Nagano, I. 1976. "New separation technique for waste plastics." *Resource Recovery and Conservation*, 2(2), pp. 127–45.

NCRR Briefs. 1980. "Refuse-derived fuel." National Center for Resource Recovery, Washington, D.C.

Newell, Scott. 1972. "Technology and economics of large shredding machines," in *Proceeding of the Third Mineral Waste Utilization Symposium*, sponsored by the U.S. Bureau of Mines and IIT Research Institute, held in Chicago, IL, March 14–16.

Office of Technology Assessment. 1979. "Economics of centralized resource recovery," in *Materials and Energy from Municipal Waste*, pp. 119–32. Washington, D.C.: Congress of the United States.

Opinion Research Corporation. 1975. "Members of special public groups evaluate plastics and their impact on the environment." December, prepared for The Society of the Plastics Industry, New York, NY.

Parker, Lorie. 1983. "Oregon's pioneering recycling act." *Environmental Law*, 15(Winter), pp. 387–411.

Paul, Bill. 1985. "Waste-to-energy plants are becoming popular with U.S. cities and towns." *Wall Street Journal*, December 16, p. 10.

Pearce, David and Richard Grace. 1976. "Stabilizing secondary materials markets." *Resources Policy*, 2(2), pp. 118–27.

Pearse, M. J. and T. J. Hickey. 1978. "Separation of mixed plastics using a dry, triboelectric technique." *Resource Recovery and Conservation*, 3(2), pp. 179–90.

Peterson, Ivars. 1984. "New life for old plastics." *Science News*, September 1, pp. 140–41.

Pieszak, Daniel A. and Kent B. Connole. 1976. "Cable works recycles polyvinyl chloride and black polyethylene." Paper presented at the 34th Annual Technical Conference of the Society of Plastics Engineers, April 26–29, Atlantic City, NJ.

Plastic Beverage Container Division. Undated. "Plastic soft drink bottle recycling." Society of the Plastics Industry, New York, NY.

Plastic Bottle Institute. 1981a. "A position paper on energy use and resource recovery." August, The Society of the Plastics Industry, New York, NY.

Plastic Bottle Institute. 1981b. "A position paper on U.S. municipal solid waste management." August, Society of the Plastics Industry, New York, NY.

Plastic Bottle Institute. 1985. "Plastic bottle recycling directory and reference guide 1985." Society of the Plastics Industry, New York, NY.

Plastic Bottle Reporter. 1983. "Tests show reclaimed PET beverage bottles recyclable into clean efficient fuel source." March, 1(2), p. 1. Plastic Bottle Information Bureau, Society of the Plastics Industry, New York, NY.

Plastic Bottle Reporter. 1984a. "PET recycler predicts more products from reclaimed plastic soft drink bottles." April 2(3), p. 1. Plastic Bottle Information Bureau, Society of the Plastics Industry, New York, NY.

Plastic Bottle Reporter. 1984b. "Reclaimed PET or HDPE used to make plastic bottle carrier." October 3(1), pp. 1–2. Plastic Bottle Information Bureau, Society of the Plastics Industry, New York; NY.

Plastic Bottle Reporter. 1985a. "PET recyclers accomplished what others viewed impossible." January 3(2), p. 1. Plastic Bottle Information Bureau, Society of the Plastics Industry, New York, NY.

Plastic Bottle Reporter. 1985b. "Plastics recycling foundation names officers; Rutgers to be research center." Summer 3(3), p. 1. Plastic Bottle Information Bureau, Society of the Plastics Industry, New York, NY.

Plastics Institute of America. Draft. "Research report — Phase II: Secondary reclamation of plastic waste." Hoboken, NJ. Prepared for Energy Conversion and Utilization Technologies Program, U.S. Department of Energy, Washington, D.C.

Plastics World. 1982. "Reclamation system grinds reinforced thermoset for reuse as filler." January, p. 86.

Plastics World. 1983. "Japanese firm finds way to recycle PET." May, p. 23.

Plastics World. 1984. "Recycler offers moldable PET." June, p. 35.

Poller, Robert C. 1980. "Reclamation of waste plastics and rubber: Recovery of materials and energy." *J. Chemical Tech. Biotech.*, 30, pp. 152–60.

Porter, Martin D. 1979. "Freight rates may discriminate against recycled materials." *Natural Res. J.*, 19(1), pp. 229–33.

Potts, J. E. 1970. "Continuous pyrolysis of plastic wastes." *Industrial Water Engineering*, August, pp. 32–35.

Prepared Foods. 1984. "Food business: Packaging news: Recycled plastic soft drink bottles used for sewing thread and boat hulls." July, p. 26.

Quade, Vicki. 1982. "Bottle bills advance . . . little by little." *American Bar Assoc. J.*, 68(September), pp. 1073–74.

Quimby, Thomas H. E. 1975. *Recycling: The Alternative to Disposal*. Washington, D.C.: Resources for the Future.

Renshaw, Edward F. 1976. "Recycling bottles." *Policy Analysis*, 2(3), pp. 493–94.

Research Triangle Institute. 1976. "Energy and economic impacts of mandatory deposits." FEA/D-76/406, Research Triangle Park, NC. Prepared for the U.S. Federal Energy Administration, Washington, D.C.

Rigby, L. J. 1981. "The collection and identification of toxic volatiles from plastics under thermal stress." *Ann. Occupational Hygiene*, 24(4), pp. 331-45.

Robert R. Nathan Associates. 1982. "Iron and steel scrap: Its accumulation and availability updated and revised to December 31, 1981." December 23, prepared for the Metal Scrap Research and Education Foundation, Washington, D.C.

Rosett, L. K. 1983. "Unique reinforcement/additive from recycled circuit boards." *Plastics Compounding*, November/December, pp. 47-54.

Roth, Terence. 1985. "Effort to pass bottle bills gets new push." *Wall Street Journal*, April 9, p. 6.

Saitoh, K., I. Nagano, and S. Izumi. 1976. "New separation technique for waste plastics." *Resource Recovery and Conservation*, 2(2), pp. 127-45.

Sattin, Bruce M. 1978. "Barriers to the use of secondary metals." Paper in the *Proceedings of the Sixth Mineral Waste Utilization Symposium*, sponsored by the U.S. Bureau of Mines and IIT Research Institute, held in Chicago, Illinois, May 2-3.

Savage, G. M., L. F. Diaz, and G. J. Trezek. 1981. "Comparative study of seven air classifiers utilized in resource recovery processing." Paper presented at the EPA 7th Annual Research Symposium, Philadelphia, March 16-18.

Savino, Anthony and Robert N. Gould. 1984. "How the nation's 15 largest cities manage solid waste." *Waste Age*, August, pp. 23-26.

Schreiber, H. P. 1977. "How will plastic waste be used in 20 years?" *Plastics Engineering*, January, pp. 40-44.

Schulz, Helmut W. 1975. "Cost/benefits of solid waste reuse." *Environmental Science and Technology*, 9(5), pp. 423-27.

Scrap Age. 1980. "Exclusive updated survey of automobile shredding." October, pp. 91-98.

Seymour, Raymond B. and Jose M. Sosa. 1977. "Plastics from plastics." *Chemtech*, August, 507-11.

Sharpe, Lois. 1977. "Public perceptions of conservation and recycling." *Materials and Society*, 1, pp. 157-61.

Sherwin, E. T. and A. R. Nollet. 1980. "Solid waste resource recovery: Technology assessment." *Mechanical Engineering*, May, pp. 26-35.

Showyin, L. 1971. "The problems associated with the disposal and/or reuse of plastic containers," in *Solid Waste Treatment and Disposal: The International Edition of the 1971 Australian Waste Disposal Conference* held at the University of New South Wales, pp. 187-89. New York: Kirov.

Smith, H. Verity. 1978. "Some criteria for the successful commercial recycling of heterogeneous plastics waste." *Conservation and Recycling*, 2(2), pp. 197-201.
_____. 1979. "The recycling of mixed thermoplastics waste." *Polymer-Plastics Technology and Engineering*, 12(2), pp. 141-47.

Smoluk, G. 1978. "Scrap recovery: Can you afford not to?" *Plastics World*, March, 32-36.

Society of the Plastics Industry. 1977, 1982, 1983, 1984, and 1985. *Facts and Figures of the U.S. Plastics Industry*, New York, NY.

Spaak, Albert. 1984. "Recycling a mixture of plastics 'A Challenge in Today's Environment'." Plastics Institute of America, Hoboken, New Jersey. Presented at the Federation of Materials Societies Conference, Bureau of Mines, U.S. Department of the Interior, held in Washington, D.C. July 17–19.

Sperber, R. J. and S. L. Rosen. 1974. "Reuse of polymer waste." *Polymer-Plastics Technology and Engineering*, 3(2), pp. 215–39.

Technomic Consultants. 1981. "Final report: Plastic recovery trends and opportunities in post-consumer segments," March 8. Prepared for the Society of thbe Plastics Industry, New York, NY.

Throne, J. L. and R. G. Griskey. 1972. "Heating values and thermochemical properties of plastics." *Modern Plastics*, November, pp. 96–100.

Tillman, David A. 1975. "Fuels from recycling systems." *Environmental Science and Technology*, 9(5), pp. 418–22.

Trauernicht, J. O. 1981. "U.S. urged to spur recycling of plastics." *J. Commerce*, May 6, p. 3.

Ulrich, Henry. 1978. "Recycling of polyurethane and isocyanurate foam." *Advances in Urethane Science and Technology*, 5, pp. 49–57.

U.S. Bureau of the Census. Various issues. *Statistical Abstracts of the United States*. U.S. Department of Commerce, Washington, D.C.: Government Printing Office.

U.S. Bureau of Economic Analysis. Various issues. *Business Statistics*. U.S. Department of Commerce, Washington, D.C.: Government Printing Office.

U.S. Bureau of Economic Analysis. Various issues. *Survey of Current Business*. U.S. Department of Commerce, Washington, D.C.: Government Printing Office.

U.S. Bureau of Labor Statistics. Various issues. *Employment and Earnings*. U.S. Department of Labor, Washington, D.C.: Government Printing Office.

U.S. Environmental Protection Agency. 1984. Personal conversation with staff members, March 26, 1984.

U.S. Department of Energy. Any recent issue. *Monthly Energy Review*. Washington, D.C.: Government Printing Office.

Valdez, E. G. 1976. "Separation of plastics from automobile scrap," in *Proceedings of the Fifth Mineral Waste Utilization Symposium*, sponsored by the U.S. Bureau of Mines and IIT Research Institute, held in Chicago, Illinois, April 13–14.

Valdez, E. G., K. C. Dean, J. H. Bilbrey, Jr., and L. R. Mahoney. 1975. "Recovering polyurethane foam and other plastics from auto-shredder reject." RI 8091, Bureau of Mines, U.S. Department of the Interior, Washington, D.C.

Vaughan, D. A., M. Y. Anastas, and H. H. Krause. 1974. "An analysis of the current impact of plastic refuse disposal upon the environment." EPA-670/2-74-083, December, Battelle Columbus Laboratories, report for the U.S. Environmental Protection Agency, Washington, D.C.

Vaughan, D. A., C. Ifeadi, R. A. Markle, and H. H. Krause. 1975. "Environmental assessment of future disposal methods for plastics in municipal solid waste." EPA-670/2-75-058, June, Battelle Columbus Laboratories, report for the U.S. Environmental Protection Agency, Washington, D.C.

Velzy, Charles O. 1985. "Measurement of dioxin emissions of energy-from-waste plants." *Waste Age*, April, pp. 186–90.

Versar, Incorporated. 1982. "Assessments of future environmental trends and problems of increased use, recycling, and combustion of fiber-reinforced, plastic and metal composite materials." EPA/600/882/019, July 14, report for the U.S. Environmental Protection Agency, Washington, D.C.

Vesilind, P. A. and J. J. Pierce. 1983. "Fundamental aspects of air classifier operation and design." DOE/CS/20544-1, January, Duke University, prepared for the U.S. Department of Energy, Washington, D.C.

Wakefield, Jeffrey M. 1980. "Problems associated with the management of solid wastes: Is there a solution in the offing?" *West Virginia Law Review*, 83(Fall), pp. 131-57.

Waste Age. 1979. "Goodyear demonstrates viability of polyester recycling." December, p. 2.

Waste Age. 1984. "Resource recovery activities report." November, pp. 91-118.

Waste Age. 1985a. "States and landfills: Solid waste survey produces a snapshot." January, pp. 37-40.

Waste Age. 1985b. "Landfill survey: Few large facilities reported." August, pp. 56-62.

Wehrenberg, Robert H., II. 1979. "Plastics recycling: Is it now commercially feasible?" *Mechanical Engineering*, March, pp. 34-39.

_____. 1980. "Graphite composites—The key to lightweight autos." *Materials Engineering*, January, pp. 36-39.

_____. 1982. "Practical plastics recycling." *Mechanical Engineering*, February, pp. 44-50.

Welsh, Richar O. and Robert G. Hunt. 1978. "Resource and environmental profile analysis of five milk container systems." EPA/530/SW-168C, Midwest Research Corporation, prepared for the U.S. Environmental Protection Agency, Washington, D.C.: U.S. Government Printing Office.

Wilson, David Gordon, and Stephen David Senturia. 1975. "Design considerations for a pilot process for separating municipal refuse." EPA/670/2-75-040, May, Massachusetts Institute of Technology, prepared for the U.S. Environmental Protection Agency, Washington, D.C.

Yacona, Terry. Undated. "Additional information on recycling plastic beverage bottles." Plastic Bottle Information Bureau, Society of the Plastics Industry, New York, NY.

Zalob, David S. 1979. "Current legislation and practice of compulsory recycling: An international perspective." *Natural Res. J.*, 19(3), pp. 611-28.

Zikmund, William G. and William J. Stanton. 1971. "Recycling solid wastes: A channels-of-distribution problem." *J. Marketing*, 35(3), pp. 34-39.

Index

About the Author

T. Randall Curlee is an economist and research staff member in the Energy and Economic Analysis Section within the Energy Division at the Oak Ridge National Laboratory. Dr. Curlee has published several articles and reports on various aspects of plastics recycling. He has also published in the areas of oil vulnerability and strategic stockpiling, oil pricing, uranium markets, and the market penetration of new materials and technologies.

Dr. Curlee received his B.S. degree in economics from the University of Tennessee in 1976. He was awarded the M.S. and Ph.D. degrees in economics from Purdue University in 1978 and 1981, respectively.